Development of a framework for participatory ergonomics

Helen M Haines and Professor John R Wilson
Institute for Occupational Ergonomics
Department of Manufacturing Engineering and
Operations Management
University of Nottingham
University Park
Nottingham
NG7 2RD

There has been a significant growth of interest in participatory ergonomics during the last ten years. Many of the rapidly increasing number of 'participatory' initiatives reported worldwide are attempting to use this kind of approach with the aim of improving the health and safety of their employees. However, it is important to recognise that what counts or is understood as being a participatory ergonomics initiative can vary widely. There would appear to be a need, therefore, to help clarify the concept, to provide some kind of definition and guidance for its use. This report reviews the available literature on participatory ergonomics and attempts to synthesise the current information. It also examines the likely costs, benefits and requirements for participatory ergonomics programmes as well as identifying possible avenues for further research. Thirdly, the report presents a preliminary framework for the implementation of different types of initiative in this fast developing field.

This report and the work it describes were funded by the Health and Safety Executive. Its contents, including any opinions and/or conclusions expressed, are those of the Authors alone and do not necessarily reflect HSE policy.

HSE BOOKS

First published 1998

ISBN 0 7176 1573 1

Contents

1.0 Introduction and aims

Organisations are increasingly appreciating the benefits associated with applying ergonomics to the design of workplaces and jobs. However, along with this increased interest in ergonomics has been the recognition that ergonomics experts alone cannot achieve widespread and successful implementation of ergonomics at work. On a practical level it has been found that there are simply not enough experts to fulfil the needs of industry. Moreover, the employment of such a person may not be cost effective for many smaller organisations. A further reason why an outside expert may be of limited value is that they might be too isolated to be the core of an ergonomics and safety culture within the company compared to the alternative, namely an internally managed participative ergonomics programme.

There has been a significant growth of interest in participatory ergonomics during the last ten years. Many of the rapidly expanding number of 'participatory' initiatives reported worldwide are attempting to use this kind of approach with the aim of improving the health and safety of their employees. However, it is important to recognise that what counts or is understood as being a participatory ergonomics initiative can vary widely. There would appear to be a need, therefore, to help clarify the concept, to provide some kind of definition and framework to guide its use. Consequently, the aims of this report are to:

1. review the available literature on participatory ergonomics and synthesise information and data as regards the state of the art
2. identify the likely costs, benefits and requirements for effective participatory ergonomics programmes
3. offer a preliminary framework which could be used as the basis for further applied research.

The project and this report have drawn upon several bodies of work. A selection of the participatory ergonomics literature is used with the specific aim of examining improvements in health and safety at work. The report also draws from a large body of work on participative practices at work which stands outside of the existing ergonomics literature. As well as published information, the report takes into consideration some of the ideas and experiences of the authors at the University of Nottingham and also those of a number of international researchers.

By including approaches both from outside ergonomics and also those initiatives where participation is part of an ergonomics contribution, the body of source information is very large. Therefore, only such ideas, opinions and findings as will provide a first structured view of the remit, value and processes of participatory ergonomics are included in the report. Some of the definitions of participatory ergonomics are examined before going on to look at a selection of models and structures reported in the literature. The context for participatory ergonomics is then examined, both in terms of trends in participatory management generally and also in terms of its relationship to health and safety at work. After this the pros and cons of taking a participatory approach to ergonomics are discussed before moving on to look at some of the tools and methods which have been used in various initiatives. Towards the latter stages of the report the issue of evaluating participatory ergonomics programmes is considered after which a brief international perspective on participatory requirements is presented. Finally, a framework for participatory ergonomics is described along with a discussion of some of the research still required in this area.

This report is written for the Health and Safety Executive and as such concentrates on the use of participatory ergonomics to improve health and safety at work. Previous work by the authors has identified five circumstances in which it is vital to understand about work organisation issues in order to improve ergonomics and health and safety at work (these have been identified primarily, but not exclusively, from the point of view of reducing musculoskeletal disorders at work). This work has now been extended to examine the role participatory ergonomics may play in reducing work related ill-health and injury. A first model, developed by the authors, is presented in figure 1 illustrating the mechanisms which may be involved.

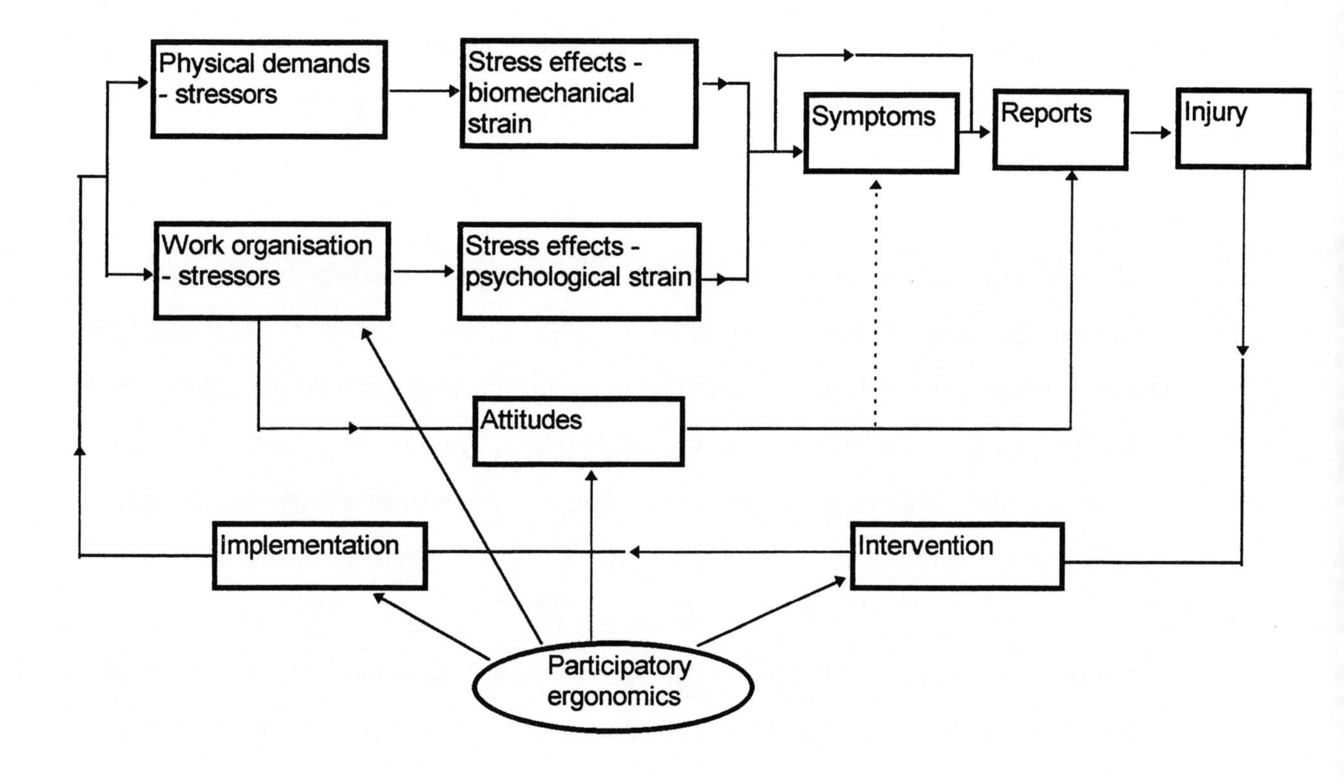

Figure 1: A model illustrating the potential role of participatory ergonomics in reducing work-related ill health and injury.

2.0 Definitions of participatory ergonomics

One of the central issues with participatory ergonomics is that there is no general agreement about what the term actually means. It has been seen by different people at different times as variously a philosophy, an approach or strategy, a programme or even a set of tools and techniques. Neither is this lack of agreement limited to participatory ergonomics; similar views have been expressed by those in the participative management field, where Cotton (1993) - talking about employee involvement - describes it as a 'fuzzy' concept (see also Marchington et al, 1994). If the existing literature in participatory ergonomics is consulted, this 'fuzziness' is easy to see. For example, Nagamachi (1995: pp 371) defines participatory ergonomics as "the workers active involvement in complementary ergonomic knowledge and procedures in their workplace ... supported by their supervisors and managers in order to improve working conditions and product quality" whereas Kourinka (1997: pp. 268) sees it as "practical ergonomics with participation of the necessary actors in problem solving".

A different view again is taken by Imada (1991); for him participatory ergonomics is one perspective in systems (or macro) ergonomics, which requires end-users as the beneficiaries of ergonomics to be involved in developing and implementing technology. This is a view shared by Lewis et al (1988: pp. 756) who state that "The rationale behind participatory ergonomics is to involve the end-user in the change process so that he/she becomes an advocate and an active change agent rather than a passive recipient of the process".

A few researchers have argued that participatory ergonomics should not be thought of as a unitary concept, but instead 'it' should be divided up in some way. Batt and Appelbaum (1995), for example make a distinction between 'off-line' and 'on-line' participation where 'off-line' participation refers to 'off the job' problem solving processes and 'on-line' participation where decision making about work is part of the job (for example through work teams). Similarly, Zink (1996) draws a distinction between "selective participation" and the attempt to achieve continuous improvement. In our own work on participatory ergonomics (Wilson and Haines, 1997) a distinction has been made between the use of participative techniques within an ergonomics investigation and intervention and the implementation of ergonomics within a participative framework or organisation.

Participatory ergonomics is clearly a complex concept involving a number of different dimensions. However, at its most basic it would seem to consist of 'stakeholders' contributing to an ergonomics initiative or sharing ergonomics knowledge and methods (where stakeholders are a broader category than 'workers', including anyone affected by the process or consequent change). For the purposes of this report, however, and reflecting the broad range of potential ergonomics initiatives across workplaces, jobs and work organisations, a first working definition is:

"The involvement of people in planning and controlling a significant amount of their own work activities, with sufficient knowledge and power to influence both processes and outcomes in order to achieve desirable goals" (Wilson, 1995a).

3.0 The context for participatory ergonomics

It is important to understand why there has been such a significant rise in interest in participatory ergonomics in recent years. Part of the explanation involves the wider social and political context. Despite some reverses and obvious exceptions, the developed world can be seen to have become more participative over the past 40 or 50 years. The public and the workforce in many countries will not accept management practices which were widespread half a century ago. Disappointment with technical investments and recognition that motivation and performance are complex issues has spawned human centred manufacturing initiatives with emphasis upon an educated, involved and responsible workforce. Today there is a greater emphasis on attributes of quality, flexibility and customer service, rather than merely quantity of output, which in turn supports the need for greater workforce participation (Karwowski and Salvendy, 1994; Salvendy and Karwowski, 1994) . High performing (or so-called 'world class') companies are said to have employee empowerment as one of the enablers to be successful, and this has often been in tandem with teamwork. "People involvement" programmes have grown out of quality circles and action groups as a part of Total Quality Management (TQM). Participative management is increasingly accepted as the effective approach to change and to organisation of work. Against this, some view TQM and its derivatives as subtle forms of management control. According to this, albeit minority, view empowerment and participation become interpreted as insidious means by which worker power is eroded in the context of industrial disputes and attempts to downsize. This argument is taken up again in section 5.0.

Within the **specific domain of health and safety**, reports of organisations using a participatory approach to tackle musculoskeletal disorders are becoming more prevalent. This is because such an approach appears to offer a number of important benefits. Involving current job holders in analysis, diagnosis and redesign can result in improved ideas and information. Because participants are, after all, those who should know most about the good and bad points of the current situation, their involvement should result in a more effective and satisfying development (although only if the conditions are established to motivate and support their contribution). Identification of such an advantage should not disguise the fact that it may be difficult for participants to express or otherwise contribute their knowledge and ideas. Sometimes people genuinely do not realise what they do or why they do it (a bottleneck also found in knowledge elicitation for 'expert systems') or else cannot find the words to describe their work and any problems or improvements to others. Involvement in participatory ergonomics may also increase acceptance of health and safety initiatives and help to open up channels of communication generally between workers and management. There is now a growing body of evidence for a link between psychosocial factors and musculoskeletal disorders (e.g. Bongers and Houtman, 1996) which suggests that the psychological implications of participation are very important. In his influential work, Robert Karasek has emphasised the actual process of participatory work as combining increases in social interaction and support with greater opportunities for decision making and control over one's work activities (e.g. Karasek and Theorell, 1990 - needless to say, this has profound implications in terms of precisely who is involved in participation - see section 10). As mentioned in the introduction, there is a need for organisations to be more self-reliant in ergonomics. Most organisations do not have the resources to have a consultant ergonomist examine every one of their work processes (even if there were enough ergonomists to go around). Moreover, such an "outside oriented" approach does not fit in with the concept of a health and safety culture of which internal ownership of ergonomics can form an important part (HSE, 1996). For a further discussion of the advantages of participatory ergonomics see section 5.0.

There have also been developments in European health and safety legislation which are relevant to participatory ergonomics. The Framework Directive (89/391/EEC) on "the introduction of measures to encourage improvements in the safety and health of workers at work" contained specific provisions about consultation and participation of workers.

Legislation in the UK now requires employers to consult employees who are not already represented by trades union safety representatives, about matters that affect their health and safety. Guidance to the Health and Safety (Consultation with Employees) Regulations 1996 states that this consultation may be carried out directly or through elected representatives and may be achieved through a variety of means such as briefing meetings, quality circles or surveys. The guidance also makes very clear the difference between informing employees and consulting them, stating that "consultation involves listening to their views and taking account of what they say before any decision is taken." (HSE, 1996; pp 5).

In the United States, guidelines and standards have had some influence on ergonomics programmes aimed at controlling the risk of musculoskeletal disorders. For instance, the Ergonomics Program Management Guidelines for Meatpacking Plants (OSHA, 1991) and the draft ANSI standard Z-365 (Control of Work-Related Cumulative Trauma Disorders) are both cases in point (see section 8.2 for further details). In the UK, British Standard BS 8800: 1996 (Occupational Health and Safety Management Systems) makes clear the requirement for employee involvement and consultation (see also HSE guidance document 'Successful health and safety management', HSE, 1993).

4.0 Models and structures of participatory ergonomics

Accepting that, for a number of reasons, there has been a rise in interest in participatory ergonomics, some of the fundamental and practical developments should be examined more closely. Having discussed some of the reasons why there has been a rise in interest in participatory ergonomics, it is time to look more closely at some of the developments themselves. Whilst Vink et al (1992) concluded that there is no single, unifying model or theoretical framework for participatory ergonomics, there has been some work in this area. Haims and Carayon (1996) for example, have produced a research model based on behavioural cybernetic principles which stresses the importance of control, action, feedback and learning. More specifically, it proposes that participants who take part in practical ergonomics training and are provided with feedback by a trainer will change their perceptions of the work environment. This increased understanding will, they suggest, pave the way for further learning and changes in participant's activities and behaviour. Furthermore, they argue that as a result of this process there will in turn be greater opportunities for further learning, action and control. It is said that this gradual development of participants over time will allow for the control of the programme to be steadily transferred from outside experts to members of the organisation, with the ultimate aim being to establish a permanent, participatory ergonomics programme capable of dealing internally with ergonomics problems.

Another contribution appears in the work of Liker et al (1989) who compared participatory ergonomics programmes in the United States and Japan. In an effort to analyse the differences between programmes they produced six models of participation where this is seen as being a combination of two separate dimensions. The first dimension (which is taken from the work of Vroom and Yetton, 1973) outlines three modes of participation: firstly, where managers make decisions having consulted with individuals; secondly where managers do the same having organised group consultations and thirdly, where managers and staff negotiate joint decisions. The other dimension is taken from the work of Coch and French (1948) who drew the distinction between "direct" and "representative" participation; in the former case the worker contributes to the consultative process and in the latter an individual is selected as a spokesperson. Liker et al's six-model framework ranges, therefore, from a situation where managers consult individual worker representatives yet still make the decisions themselves (model I) to one where all employees both participate in consultations and have a role in the decision making process (model VI).

Several other models have been advanced in the participative management field, although these also vary quite considerably depending on what view of the participatory process is being taken. Cotton (1993) describes how, in developing their models, researchers have differed in the definitions of participation they have used, the theoretical perspectives they have taken and the outcomes they have examined.

Turning to the practice of participatory ergonomics, a number of different participatory structures can be identified from the initiatives reported in the literature. For instance, some organisation-wide programmes to implement ergonomics are structured around one or more multifunction teams or 'task forces' (e.g. see Jones, 1997). These are generally made up of representatives from management, workers and technical staff such as engineers or health and safety personnel. An ergonomist may provide 'expert' ergonomics input, often by providing initial training and then by guiding the work of the team (see section 11.3 for a discussion of the changing role of the ergonomist). Although participatory ergonomics may be structured around a team, task force or committee, input from other members of the organisation may often be sought. This can be done either directly or through the use of suggestion schemes or questionnaires. For example, St Vincent et al (forthcoming b) describe a case where the permanent ergonomics committee was supplemented by both the foreman and one or two worker representatives from each work area under investigation.

An alternative participatory structure is described by Zink (1996) who looked at the implementation of a new CAD system. Because this change affected workers across the organisation, support and co-operation from all departments was essential. Consequently, participation was focused upon a number of small core teams each containing a facilitator who in turn sat as part of a project team led by a project manager. The project manager then reported to a steering committee made up of works council and higher management representatives. For Zink this mixture of 'top down' and 'bottom up' approaches was essential for efficient planning and implementation of a project of this size and complexity.

Participatory ergonomics may be integrated into participative management structures already in place within an organisation. For example, a number of studies report how quality circles or safety circles assume some responsibilities for ergonomics. In the Japanese cases reported by Liker et al (1989) quality circles provided ergonomics data and suggestions for improvement. Staff members (e.g. industrial engineers, production supervisors) were responsible for some of the analysis work, equipment design and training whilst most 'final decisions' remained the province of the steering committee.

Semi autonomous work groups are another example of a participative structure which can incorporate ergonomics. For example, Pransky et al (1996) described how, when the work organisation of a paper products manufacturer was reorganised, high performance teams were installed each of which was expected to address any problems in their work area related to quality, productivity and ergonomics. In addition, all of the teams elected someone to take part in the plant's new ergonomics group.

It is also possible to find many examples of small numbers of workers being formed into groups to tackle particular ergonomics projects. In some cases these have involved a group of employees analysing and redesigning their own work environment either as a 'one-off' initiative or as part of ongoing company-wide ergonomics programmes (see, for example, Moore, 1994; Laitinen et al, 1997) Alternatively, the remit of the project has been to design a piece of work equipment or some other kind of 'product'. The workers that are involved in this situation will not necessarily be the eventual users of the equipment, instead their role may be to provide insight and information in order to improve its design. One example of this is to be found in the work of Carter and Boquist (1995) who describe a participatory project to design a mechanical aid for concrete tile laying. In addition to the ergonomists, the project group consisted of tile layers, foremen, representatives from management, contractors, representatives of a tile manufacturer and an engineer from the company who would produce the aid. The authors explain that this particular group was chosen to represent all the parties who would have an interest in the product (i.e. all the stakeholders).

In summary, it is clear that employee participation in general, and participation in ergonomics initiatives in particular, can vary widely. One question which remains is which kind of participation is appropriate in any given context ? Cohen (1996) suggests that a number of so called 'shaping factors' should be considered in making such decisions namely,

- the nature of the issues requiring consideration
- whether the matters are broad-based or specific to a local operation or group
- whether the needs for response or action are time-limited or necessitate continuing efforts
- the abilities of the group most affected
- the organisation's prevailing practices for joint labour-management or participative approaches in resolving workplace issues.

As such, Cohen's proposals fit in with the "contingency" model of participation (Brown, 1990; Gjessing et al 1994) which stresses that because no single approach will be effective in all situations the role of the ergonomist is to encourage management to utilise the most appropriate strategy.

5.0 The pros and cons of participatory ergonomics

It should be clear that a participatory approach to ergonomics can have important advantages or strengths. However, it is not necessarily the case that it is always the best route to take. Both the advantages and disadvantages of participatory ergonomics must be examined so that informed judgements can be made concerning its utilisation. Before doing so it should be noted that information about the pros and cons of participatory ergonomics does not, in the main, come from systematic studies of the various factors involved. As Gjessing et al (1994) point out, very few reports analyse the workings of participatory ergonomics in causal terms, due, in no small part to the problems of trying to isolate variables and the size of effort in setting up experimental and control conditions (see also Buckle and Li, 1996).

Of all the advantages of participatory ergonomics, there are two direct benefits that are commonly referred to in the literature. Firstly there is the point that employees have unique knowledge and experience of work. Their involvement should, therefore, provide a clearer understanding of both the types of problems being encountered and the solutions that will be appropriate. Secondly, involving people in analysis, development and implementation of a change should generate greater feelings of solution ownership and thus may breed a greater commitment to the changes being implemented (e.g. Imada and Robertson, 1987; Wilson, 1995b).

Related to these advantages, participation may often be a learning experience for those involved. For workers, knowledge of their own work and organisation may be increased (e.g. see Buckle and Ray, 1991) Involvement in a development or implementation process can mean faster and deeper learning of a new system and hence decreased training costs and improved performance (e.g. see Wilson and Grey Taylor, 1995; Kuorinka and Patry; 1995). Participatory ergonomics may also be a learning experience for designers or other technical specialists. According to Sanoff (1985) the very departure from conventional thinking can improve design effectiveness. Similarly, St Vincent et al (forthcoming b) describe how adopting a participatory approach meant that engineers in an electrical assembly plant were able to take a more realistic view of making changes to the work.

Widening the scope of potential benefits, participatory ergonomics can, if successful, sow the seeds for its own extension. By spreading interest, understanding and expertise in ergonomics amongst those involved, and by attracting interest and desire for involvement amongst work colleagues, solutions can be generalised and ergonomics transferred elsewhere throughout an organisation (Daniellou and Garrigou, 1992; Daniellou et al, 1990). Certainly if people are involved in making ergonomics changes at work, they are more likely to be able to adapt when circumstances change or to use the ideas elsewhere, compared to where the solution was developed and installed by an outside agency. Imada (1991) refers to this as creating a 'flexible problem solving tool'. Similarly, Buckle (1996) describes how participatory ergonomics can create an increased 'in-house knowledge base' of ergonomics.

The very process of participation may provide benefits for the individuals concerned. For example, Mambrey et al (1987) found that people who participate directly show more self-confidence, competence and independence and attach more importance to self-determination than their colleagues. Such attributes may, of course, be seen as the *causes* rather than the *effects* of participation, but Mambrey and his colleagues suggest otherwise. Eklund (forthcoming) describes how worker participation in the entire problem solving process can increase opportunities for personal development, social contact, feedback, influence, challenge and variability, the very characteristics which have been identified as contributing towards 'good' work (see Karasek and Theorell, 1990; Wall et al, 1990 and Hackman and Oldman, 1980). In addition, St Vincent et al (forthcoming a) describe how having their opinions listened to increased operator's perceptions of their personal value. It has also been suggested that worker participation can play a part in reducing stress at work. Smith and Zehel (1992) summarise the so-called balance theory of job design (see Smith and Carayon-Sainfort, 1989). In this model the workplace is divided into five dimensions - the environment, the task, the technology being used, the organisational context and the person. Each of these dimensions can have a negative or positive influence but the idea is that the overall system must be 'balanced' in order to reduce stress. In terms of worker participation, for example, allowing workers to be involved in decisions about their work can balance the negative influences of low job content.

The effects of worker participation on satisfaction and productivity have been the subject of debate in the participative management literature (e.g. Cotton et al, 1988; Leana et al, 1990). A variety of quite different forms of participation have been examined (for example, gainsharing plans, self-directed work teams and quality circles amongst others). In an attempt to reach a general conclusion, Wagner's (1994) meta-analysis suggests that participation can have significant (albeit small) effects on performance and satisfaction (although Cotton, 1993, argues that because different forms of employee involvement can have very different effects, such analyses may, in fact, underestimate the potential benefits). Overall, many commentators feel that the greatest benefits come from the more extensive forms of participation (see Glew et al, 1995; Neumann, 1989 and Batt and Applebaum, 1995).

Finally, the relationship between participative processes and groups or teams is also relevant here. Basic workplace ergonomics and health and safety interventions can be carried out very efficiently and effectively through work teams participating together to identify problems and implement improvements. Moreover, involvement in such initiatives can help give the work teams a specific focus and strengthen the team basis and functioning. Group working anyway is a highly appropriate mechanism through which to enrich job designs and reduce job strain via social interaction and support (Karasek and Theorell, 1990; Wall et al, 1990). Furthermore, the socio-technical principle of compatibility (Cherns, 1976; 1987) dictates that it is desirable for the groups that develop ideas for new workplaces, work tasks and work structures to subsequently form into work teams themselves and *vice versa.*

Despite the considerable potential benefits of participatory ergonomics, it can also bring with it some problems. To begin with, it is not always that easy to instigate or support. One of the main obstacles here is people's (un)willingness to get involved. It could be that various groups or individuals are resistant. Management, for example, might see participation as a threat to their right to manage rather than as an aid (Mumford, 1991). Alternatively individual workers might feel that what they are being asked to do is outside their capabilities or influence ('cognitive limits' according to Bernoux, 1994). They might also refuse because they lack, or feel that they lack, sufficient motivation, time, and energy. Finally, workers might defer too much either to management or to technical 'experts' (Mumford, 1991) especially if they believe that they have not themselves had sufficient technical training (Garrigou et al, 1995).

However, Neumann (1989) prefers to look for explanations of why people don't participate in the structure and environment of participation itself rather than from the personality or attitudes of people. Thus she identifies *structured, relational* and *societal* explanations for non-participation. Structural explanations are those that suggest that the organisation is not suited to participation - real decisions are taken elsewhere, there is no support, or reinforcement or encouragement from job designs, training or reward systems. Relational explanations include poor or discouraging management, strictly hierarchical structures and conflicts with life outside work. Societal explanations consist of the idea that participation at work might challenge deeply held social, cultural or personal beliefs. In conclusion, Neumann suggests that we can neither expect that everyone will be willing to participate nor should we assume that any unwillingness to participate is due to some shortcoming on behalf of the individual. Rather she argues that "The choice to engage in managerial initiatives towards increased employee involvement reflects a complex calculation, conscious or otherwise, on the part of the employee." (p.8).

At the same time there are a number of other potential problems associated with the very *process* of participation. For instance, an argument may be raised against the group-based nature of participation. Indeed, group decision making has often been seen as fairly unreliable (see Fraser and Foster, 1982), often degenerating to the lowest common denominator (although it could be said that this merely emphasises the importance of appropriate structures and methods for group decision making). Secondly, it is also the case that planning and developing new systems, workplaces or organizations participatively may be slower, more complex and require greater effort than under normal circumstances. However, this may not be a problem if the resulting participative development is more appropriate. This point is illustrated in a project to implement a new CAD system which provided an opportunity to compare a participatory with a non-participatory approach. A requirement analysis was undertaken firstly by experts and then by a range of user representatives. Although the participatory workshops were more time consuming, the broad perspective they provided for the requirement analysis meant that implementation of the new technology could be planned much more precisely and the risk of failure was reduced. The author concluded that these benefits certainly offset the increased time requirement of taking a participatory approach (Zink, 1996).

Another potential problem with taking a participatory approach is that the process may encourage employees to develop unrealistic expectations. These expectations may concern the timescale for management approval, implementation and benefits from their recommendations (Cohen, 1996) or alternatively the extent to which jobs or work equipment will be changed. However, it is not always the case that employees' expectations will be unreasonable. In a report on the ergonomics programme at the Library of Congress, the authors reported quite the opposite from those involved (Mansfield and Armstrong, 1997).

There are some cases that have revealed that the process or content of participation impacts on other parts of the organisation in unhelpful ways. For example, increasing the variety of work done by word processing staff may reduce the need for other clerical personnel. A participative process that improves the workplace for some may make those in other departments envious or dissatisfied. Indeed, there may be unfortunate systemic effects from the very process of participation, in that other groups may wish to be similarly involved and this may be neither possible nor even desirable. Whether these 'ripple effects' in fact turn out to be damaging may depend upon the way in which companies deal with them. Cohen (1996) cites a case reported by Peters (1989) showing how efforts to expand self-regulated teams in the mining industry failed. Despite positive gains in safety (the very nature of the autonomous team made each worker responsible for maintaining safe conditions) the programme was rejected by the union on the grounds that the special treatment of the pilot group created an 'elitist attitude' that was resented by other miners. Westlander et al (1995) also describe a case showing how perceptions of a participatory group as 'privileged' had negative effects on the outcomes of the project.

It must also be acknowledged that some projects referred to as participatory ergonomics are so in name only. Unlike genuine initiatives, they might be little more than exercises in information provision or even manipulative programmes with a covert agenda (see Reuter, 1987 - who also quotes Arnstein, 1969 - for a typology of participation). Cynical use of participative methods may yield some limited dividends in the short term but will not provide any of the potential greater long term benefits. Such use may partially explain participation's ambiguous tradition within industrial relations as identified by Forrester (1986) in which he sees both the criteria and context of participation as being viewed with suspicion by trades unions (citing the case of quality circles which are often viewed as being used by management to by-pass procedural channels and so weakening trade unionism).

There is, finally, one more general area of difficulty when it comes to establishing proof of results. In common with many work design and ergonomics initiatives, it may be hard to show that participation has truly brought about a better system or system change, or that more autocratic methods would have been less effective (see also Shipley, 1990). Such lack of evidence, compounded by the paucity of good evaluated case studies, may contribute to a lack of face validity in the eyes of management.

Clearly a participatory approach to ergonomics problems is not a panacea. Some organisations are simply not ready for participative practices. Childs (1984), for example, identified circumstances where participation was seen as a waste of time and others where it was used to actually *obstruct* change. Yet, whilst we cannot just uncritically accept participation as appropriate in every case, as long as we give careful consideration to the time, place, the individuals involved, as well as to the structure and methods being used, then the full benefits of participation may be realised.

6.0 Participatory tools and methods

A variety of different tools and methods have been reported as useful within participatory ergonomics initiatives. Some have been borrowed, adapted or developed with participatory ergonomics specifically in mind, whereas others have been taken from more 'traditional' ergonomics initiatives and then applied within a participative exercise. The aim of this section is to discuss some of the issues associated with the development, selection and use of participative methods as well as to provide some examples from the literature.

It is clear that the variety of ways in which participatory ergonomics may be applied means that certain techniques will be much better suited to certain situations. For example, if we are looking to instigate and support participatory ergonomics as a macroergonomic strategy, we are likely to be interested in tools to sell a participative approach to stakeholders (Kuorinka, 1997) or to facilitate groupworking and improve interpersonal skills. Caccamise (1995) provides an example of this last scenario in a project (which started as a 'micro' level iniative) to redesign a job aid in the nuclear industry. An important part of this process was the need for team building. Teams were composed of personnel from quite different disciplines and these differences seemed to be associated with difficulties in working together. In an attempt to address this issue, members were given the Myers-Briggs Type Inventory as a means of identifying 'thinking styles'. The author reports that participants found the exercise useful and that the results were used successfully to identify strategies for dealing with some of the communication problems. Table 1 summarises methods or techniques used to set up or support the participatory *process*.

Method or Technique	Primary Purpose	Reference (for description or use of technique or method)
Task forces	Participation organisation	Liker et al, 1991; Wilson, 1994
Change management process	Participation organisation	Buchanan and Boddy, 1992
Team formation, building	Participation organisation	Caccamise, 1995; West, 1994
Team training	Preparation and support	Gjessing et al, 1994
"Train-the-trainers"	Preparation and support	Corlett, 1991; Silverstein et al, 1991
Stakeholder analysis	Preparation and support	Burgoyne, 1994; West, 1994

Table 1: Methods used to set-up or support the participatory process

An obvious issue in the selection and use of appropriate tools and techniques concerns participants' 'expertise'. Depending on the structure chosen for the participatory initiative, experts may play a number of different roles. They may restrict their intervention to largely supporting or guiding the participatory process, or they may be involved with a variety of other workers as members of a multifunction group. Alternatively, at the more 'hands-on' end of the scale they may undertake much of the analysis themselves, maybe through the use of a research team. An example of the latter is given by Kuorinka and Patry (1995) where, in a study prompted by an increase in the incidence of work related upper limb disorders in a poultry processing plant, analysis of the problem was undertaken by a project group. According to the authors, the participatory group's role was primarily the validation and implementation of the project group's proposals. Clearly the techniques used in this instance may vary from those we would teach to non-expert participants in order for them to analyse workplaces, generate new ideas and evaluate changes for themselves.

Nagamachi (1995) has described a number of ergonomics tools developed specifically to allow members of quality circles to analyse workplaces. These include a device designed to identify the factors necessary for job redesign (labelled a JDLC chart - Job (Re)design for Life Cycle) and a 2-D mannequin to assist in the selection of an appropriate work height. He has also developed a tool which allows working postures to be analysed on the shop floor using a pocket computer where an observer presses the relevant buttons in response to the postures observed. At the end of the observation period the computer will analyse the results and present them in terms of cumulative time spent in each posture, average relative metabolic ratio and a working posture index (Nagamachi and Tanaka, 1995).

The issue of expertise brings us on to the subject of training requirements. Many authors have highlighted the need for those involved in participatory initiatives to be provided with training in ergonomics techniques (as well as general ergonomics awareness and interpersonal skills - see section 11.3). Such training may well provide a common-ground to help participants from different occupational groups to communicate with each other (e.g. see Kuorinka, 1997). If ergonomics techniques are complicated then the training required may well be time-consuming and companies must be made aware of this, yet it can still be the case that the costs of training may be more than offset by savings made. For example, Kuorinka and Patry (1995) describe how a mixed group of workers, engineers and occupational health representatives investigated a particularly hazardous task within a steel-making plant (replacing the fire-resistant lining of processing vessels). A good deal of time and effort was spent not only teaching the group motion-time-method analysis (so that they could look at existing work methods and plan new ones) but also on team building - investments which the management considered justified by the short run-in time of the redesigned process.

By and large, however, it may be beneficial to avoid using overly complex or technical analytical tools, for as Liker et al (1989) point out, the ability of organisations to remain self-sufficient goes down as the sophistication of ergonomics analysis tools rises. Certainly, if one of the aims of participatory ergonomics is to free institutions from reliance on the outside expert, the use of highly complex tools may prove self-defeating. This may restrict the use of some computer-based modelling and simulation tools for the time being (the sophistication here lying not so much in their use as in the interpretation of the results they produce).

Tables 2 and 3 - at the end of this section - attempt to classify tools and methods according to their main uses, such as in problem analysis, idea generation or concept development. Table 2 summarises methods developed with participatory ergonomics specifically in mind, whereas Table 3 summarises methods and techniques used in 'traditional' ergonomics initiatives which have been applied within participatory initiatives.

Problem analysis tools such as pareto analyses, cause-and-effect diagrams (or 'fishbone charts') and link analyses are described by Imada (1991). He also classifies checklists amongst his description of 'participatory methods'. The advantage of checklists is that they can help participants approach problem identification in a structured way. Support for this is provided by Keyserling and Hankins (1994) who compared two approaches to analysing jobs undertaken by ergonomics committees in the automobile industry. By comparing a subset of the committee's findings with those of 'experts' they found that the group using a structured checklist were more sensitive to ergonomics 'stresses' than the experts - whereas the opposite was true for a group who adopted an 'open ended' approach. Support for non-experts using a formal analysis scheme is also given by St Vincent et al (forthcoming a).

Videorecordings can also be extremely useful when workplace problems are being analysed. Kuorinka (1997) describes the so-called autoconfrontation method. Here a videotape of a work process is shown to the group whilst the worker describes the action. Whilst, of course, the level of the narrative will vary depending on the particular study, the method is still an effective way of getting a better understanding of the problem under examination.

Brainstorming and group discussions are widely used for problem analysis as well as for generating new ideas. Caplan (1990) describes how focus groups (an approach commonly used in market research) can be used to address ergonomics problems. In a typical case, a small number of selected participants discuss a set of specific issues, facilitated by a moderator. Caplan describes how this approach can be particularly useful in coming up with solutions for problems in workplace or product design. Techniques such as word maps (see Wilson 1991a) are also a useful way of focusing group discussions.

It is important to find an adequate method of recording the salient points or decisions reached during participants' discussions. This may involve nominating 'notetakers' or using video or audiotape recordings. An alternative to this comes in the form of Frei et al's (1993) 'metaplan' technique. Here large pin boards are covered with paper on which smaller cards of different colours are pinned. Any ideas and problems are listed and regrouped on the board according to a particular theme or category. After each meeting the cards are photographed in their final position with photocopies being distributed to the group members.

The need for participants to be able to visualise both problems and solutions has also been discussed by a number of other authors. For example, small scale layout modelling may be used. Alternatively (and particularly at the evaluation stage) mock-ups of the work area may be created using easily manipulated material such as cardboard, polystyrene or foam. Thirdly, computer aided design can be used for similar purposes. One piece of research which combined a number of visualisation tools was Snow et al's (1996) work on designing a research laboratory. In the preliminary stages of the project CAD drawings were used in conjunction with 'scenario driven discussions' to allow participants to generate ideas and information and to evaluate the implications of different alternatives. Then overhead projections of these drawings were used in a 'dynamic space manipulation' exercise where coloured cutouts representing rooms and equipment were placed and repositioned on the projected images. Finally, during the later stages of the design process a 'virtual' mock-up was used which allowed users to visualise the design in three dimensions and to tour the workspace at both standing and wheelchair height. It is possible that advances in virtual reality technology may mean that virtual environments (VEs) become a viable and useful mechanism to drive relevant participatory exercises. Although head mounted displays might be used, VEs in desktop systems are technically and financially more feasible at the moment; as speeds improve, real-time networking amongst remote participants becomes a better prospect. The communication, visualisation, interactivity and 'walkthrough' capabilities of VR/VE mean it is often viewed as a tool to aid or drive collaborative exercises - for instance in simultaneous engineering (design for manufacturability/maintainability see Cobb et al, 1995) - and it is also a medium free of the conventions and mysteries of the profession or discipline (think of architect's or engineering drawings). Interfaces within the VE can be built that allow groups of users to rapidly reconfigure layouts of workstations for instance (Wilson, 1997). Certainly there will be growth in the use of VR/VE for 'customer specified' design (Kansei Engineering) and participatory redesign - of equipment, workplaces and buildings - is a short step from this.

Many of the techniques already discussed - world maps, round-robin questionnaires, layout modelling and mock-ups - also form part of Design Decision Groups (Wilson, 1991a). This concept is derived from the work of O'Brien (1981), who adapted theories and techniques from market research and the literature on creativity and innovation. O'Brien encouraged participants to use a range of "thinking tools" aimed at getting them to visualise their ideas.

All of these tools or techniques form part of what he called "Shared Experience Events". Activities included splitting groups into discussants and listeners (where the latter sub-group provide feedback only when the first group has finished), the use of dynamic agendas and the use of drawing materials to facilitate concept development and the communication of ideas. Design Decision Groups have been used in a number of applications including the design of retail checkouts and library issue desks (see Wilson, 1991a for a full explanation of the process). Lehtela and Kukkonen (1991) have subsequently used a modified version of the DDG in designing a telephone exchange where, in particular, the benefits of "reference visits" by some of the participants to other exchanges using different systems was highlighted. They suggest that people tend to base designs largely on their own experience and that an awareness of solutions adopted in other situations serves to broaden their outlook. DDG's have also been utilised as part of a much wider systematic approach to engineering and design throughout an organisation (Aikin, Rollings and Wilson, 1994; Sullivan and Wilson, 1994).

A frequent criticism of DDGs is that they are used only with white collar, professional or technical staff. However, in one study an adapted DDG (Problem Solving Group Procedure) was used with a group of crane drivers working at an incinerator plant. In this instance, problems with the design of the workplace were addressed by the drivers themselves and they also sourced and costed their solutions within a budget set by management. Furthermore, this 'one-off' initiative was extended into a process of continual improvement (Wilson, 1995b).

Finally, the use of role playing techniques and simulation games should be considered. According to Saunders (1995: pp 13) "a simulation game models those critical features of reality which involve rules, strategies and tactics" where decisions can be made in a 'risk-free environment'. Interestingly, Saunders points out that role play forms an important part of many simulation games (although here, rather than looking at the situation from their own point of view, participants are asked to employ the 'expectations' of another). In the preface to a symposium on simulation games, Vartiainen and Smeds (1995) describe how there is increasing interest in the place of simulation games within organisational development, particularly from the perspective of developing work processes or group interaction and teamwork (see also Ruohomaki, 1995; Vartiainen and Ruohamaki, 1995). It is suggested that simulation games provide a way of predicting and adapting to the rapid changes faced by many organisations.

Method or Technique	Primary Purpose	Reference (for description or use of technique or method)
"Statistical analysis measurements" (SAM)	Problem analysis	Liker et al, 1991
JDLC chart	Problem analysis	Nagamachi, 1991
Posture analysis tool	Problem analysis	Nagamachi, and Tanaka, 1995; Nagamachi, 1991
Pareto analysis	Problem analysis	Imada, 1991
Cause-and-effect diagram	Problem analysis	Imada, 1991
"Five ergonomic viewpoints"	Problem analysis	Noro, 1991
Autoconfrontation	Problem analysis	Kuorinka, 1997
Activity analysis	Problem analysis and situation prediction	Garrigou et al, 1995
Round robin questionnaire	Creativity stimulation and idea generation	O'Brien, 1981; Wilson, 1991a
Word map	Creativity stimulation and idea generation	O'Brien, 1981; Wilson, 1991a
Silent drawing/assessment	Creativity stimulation and idea generation	O'Brien, 1981; Wilson, 1991a
Shared Experience Events	Idea generation	O'Brien, 1981
Scenario driven discussions	idea generation and concept development	Snow et al, 1996
Dynamic space manipulation	idea generation and concept development	Snow et al, 1996
Role playing techniques and simulation games	Idea generation and concept evaluation	Ruohomaki, 1995; Vartiainen, 1995
Design Decision Group	Idea generation and concept evaluation	Wilson, 1991a
Problem Solving Group	Idea generation and concept evaluation	Wilson, 1995b
Layout modelling and mock-ups	Concept evaluation	Wilson, 1991a
Intervention ideas	Concept evaluation	Moir and Bucholz, 1996
'Metaplan'	Process recording	Frei et al, 1993

Table 2: Methods developed for participatory ergonomics

Method or Technique	Primary Purpose	Reference (for description of use of technique or method)
Task analysis, functional task decomposition	Problem analysis	Kirwan and Ainsworth, 1992; McNeese et al, 1995
Work study techniques (e.g. MTMM)	Problem analysis	Kuorinka and Patry, 1995; Noro, 1991
Link analysis	Problem analysis	Imada, 1991; Kirwan and Ainsworth, 1992
Brainstorming techniques	Idea generation and concept development	West, 1994
Focus groups	Idea generation and concept development	Caplan, 1990
Delphi technique	Idea generation and concept evaluation	Linstone and Turoff, 1975
Virtual reality	Concept evaluation	Snow et al, 1996
Interviews, questionnaires	Problem analysis, idea generation	Oppenheim, 1992
Checklists	Problem analysis and concept evaluation	Rawling, 1991; Meister, 1984; Sinclair, 1995

Table 3: Methods used within participatory ergonomics initiatives

7.0 Evaluation of participatory programmes

Mention has already been made of the difficulties involved in proving the effectiveness of participatory projects. Nevertheless there are a number of ways in which a participatory programme may be evaluated. The most obvious of these consists of measuring the extent to which the original, anticipated outcomes have been achieved. For example, is the incidence of musculoskeletal disorders falling, are employees more satisfied with their work, has productivity increased ? There are, however, potential difficulties with using such outcome measures. For example, in the case of musculoskeletal disorders, it may take some time for any reductions to show (Buckle and Li, 1996). Alternatively, effects might be significant but small, thereby requiring large-scale applications before statistical checks become meaningful (Cohen, 1996). Having said that, outcome measures can be of use for redesign exercises where the solutions produced may be evaluated in terms of how well they address key problems. It should be remembered, however, that as with all practical ergonomics programmes applied within an organisation, evaluation in any meaningful sense is often confounded by changes in company environment, market, structure, processes or people during the life of the implementation and evaluation.

There are a number of alternative approaches to evaluation besides the use of outcome measures, one of which is advocated by Buckle and Li (1996). Their suggestion is that a reduction in risk factors can be used as a valid measure of the effectiveness of an ergonomics programme. Yet another approach is to use 'process measures'. These look at a range of different aspects of the participation process such as the number of changes implemented, participant's satisfaction with their involvement or the spread of the programme throughout an organisation (see for example, Vink et al, 1995).

For many managers the success of a participatory ergonomics programme will be measured in terms of cost savings. However, cost-benefit analyses of ergonomics projects are not always easy to calculate. Accurate economic estimates for all factors are often very difficult to produce and some positive outcomes may not have direct economic consequences (the issues surrounding ergonomics cost-benefit analyses are discussed by a number of authors including Corlett, 1988; Simpson and Mason, 1995; Oxenburgh, 1991 and Rouse and Boff, 1997). It is not that surprising, therefore, that very few cases in the participatory ergonomics literature evaluate their success in financial terms. However, of the ones that have, several are discussed at a conference jointly organised by the University of California, Berkeley and the University of Michigan during 1995 (details can be found in the American Industrial Hygiene Association Journal, February 1997). All these cases feature company-wide ergonomics programmes incorporating some degree of worker participation, where the primary aim is to reduce musculoskeletal disorders and attempts have been made to calculate the resulting benefits.

One of the cases looks at some of the costs associated with setting up an ergonomics programme suggesting that, over a two year period, almost $108,000 was spent on training and consultation. Significantly more ($510,000) was spent on ergonomics interventions.

Although this may appear to be a large sum of money, the authors note that during the same two year period over three times that amount was spent on worker compensation costs (Mansfield and Armstrong, 1997).

Another case, this time in a poultry processing company, compared workers' compensation claims and costs over a five year period. It was found that after an initial rise, both the incidence and the severity rate (cost) of new claims for upper extremity musculoskeletal disorders had decreased significantly. This was attributed to increased awareness and early reporting of these conditions. It is important to note, however, that these results represent company averages and a more detailed analysis revealed that improvements were uneven. Indeed a few plants showed increases in claims. A number of different reasons are given to explain these variations (Jones, 1997).

Another, even more complicated set of outcomes is discussed in a report by Moore (1997). The study looked at the effects of an ergonomics programme on a number of dimensions over a six year period. During this time it was found that workers compensation costs decreased significantly. Yet paradoxically, it was also found that both crude incidence rate (for all injuries and illness) and lost-time incidence rate also increased during that time. In addition, there was no discernable pattern to changes in severity rate and no change in the percentage of recordable injuries which were classified as 'ergonomics-related'.

The final case comes from a collection of examples reported by Hendrick (1996). Structured almost like a controlled experiment, it featured seven companies that received training on how to set up a participatory programme and where all but one went on to implement them. Eighteen months later the six companies could report a total saving of over one and a quarter million dollars due to reductions in 'strain-type injuries' whereas the seventh company reported an increase in injury levels.

8.0 Developments in participatory ergonomics within different countries

In order to place the review of participatory ergonomics in a more global context, visits were made to a number of organisations and individual experts in other countries. Investigation focused on specific legislative requirements and developments within two main centres; namely Scandinavia and North America. One of the things that characterises both of these regions of the world is that they are particularly 'active' within the domain of participatory ergonomics. Hence they provide a particularly useful insight into these issues.

Visits were made to the Technical University of Denmark and the Universities of Lund and Linköping in Sweden. Within North America, visits were made to NIOSH, University of Southern California and University of Michigan. In addition factory visits were made to the Ford Motor Company in Detroit and the Nummi (Toyota and General Motors) plant in Freemont, California.

8.1 Scandinavia

Participatory ergonomics is relatively well accepted in Scandinavia. Not only have the Scandinavian countries got a long tradition of collaboration between workers and management, but work environment issues and participative management techniques of teamwork and empowerment have a strong basis in Scandinavian societies. One of the contributing factors here is that union membership tends to be high in a context where the unions, instead of being adversarial in nature, play an important co-operative role in the labour market.

Worker-management collaboration is supported by various pieces of legislation and joint agreements. Looking at the specific domain of participatory ergonomics, there is legislation and public policy which may well be relevant here. Firstly, there is a requirement for representative participation on safety matters. Here joint safety committees are established within organisations. According to Jensen (1997) such formal structures for indirect participation can form a solid basis for the introduction of more direct participatory ergonomics strategies. Norway and Sweden have introduced legislation on internal control of the work environment. This requires a formalised management system to be established. Work assessments required as part of this approach may well involve participation between managers, supervisors, operators and production engineers (Eklund, 1997). Danish legislation also requires workplace assessments to be carried out, in this case to fulfil the requirements of the EC Framework Directive. It seems to have been generally accepted within Denmark that workplace assessments should be based on a participative rather than an expert approach (Jensen, 1997).

Looking beyond legislation, a joint agreement which may be of relevance to participatory ergonomics is the Danish Action plan on the reduction of monotonous repetitive work. This 1993 agreement, signed by unions, employers and the government, has set a target for a 50% reduction in this type of work by the year 2000. Both Sweden and Denmark have occupational health services which are fairly strong in the area of participatory ergonomics (the occupational health service may be external to a company or, within larger organisations it may well be an internal structure, however it will retain some independence from the company). The Danish Occupational Health Service is dominated by physiotherapists and technical staff rather than by medical doctors. As an institution they promote the use of participative approaches and this has general support from the government and labour market organisations (Jensen, 1997). Finally, the Swedish Foundation for Working Life Fund should be mentioned. The large programme of workplace development projects it supported was largely undertaken during the first half of the 1990s. Many of these projects covered ergonomics issues including musculoskeletal disorders, organisational arrangements and participation.

Overall it is clear that the 'Scandinavian model' of collaboration is very supportive of participatory ergonomics. However, there are still important issues of debate within this area. For example, there seems to be interest in the power relationship between different people within the participatory ergonomics process (for example, between managers and workers) and in the application of the concept of 'local theory' which concerns efforts to equalise these differentials (local theory refers to the establishment of an employee based common understanding of the causes of workplace problems and the 'legitimate' basis of actions to deal with these - Jensen, 1996).

8.2 North America

One of the important issues for participatory ergonomics in the USA is the question of what is permissible in terms of labour relations legislation. For example, it would appear that the National Labor Relations Act in the States prohibits employers from organising certain 'employee involvement' groups because they are perceived to be 'sham unions' (that is, unrepresentative, company controlled bodies which are, therefore, unauthorised to make decisions on issues to do with conditions of work). This is not to say, however, that groups of this kind are not used in non-unionised companies, it is just that the rules and regulations are both complicated and contentious. What is clear though, is that in companies where workers are represented by unions, employment involvement schemes for ergonomics programmes are quite common. What is more, it would seem that these schemes are seen as being extremely worthwhile by most involved (Gjessing, 1997).

A large number of workplace ergonomics programmes have focused on addressing musculoskeletal disorders (for examples, see Cohen, 1997; Mansfield and Armstrong, 1997), reflecting the high incidence of these problems and the costs of related compensation (apparently, reducing work related musculoskeletal disorders is now on the agenda for the Clinton administration - Hendrick, 1997). In January 1997 a joint OSHA - NIOSH conference was planned in order to look at a range of these programmes. Over 70 case studies were heard from a wide variety of different organisations and one of the clearest themes or messages to come out of this gathering was that employee involvement was "essential" (Gjessing, 1997).

At present, the US government has no specific public policy or regulations on worker involvement, as it remains a bone of political contention. For example, the OSHA workplace ergonomics standard, which does encourage worker participation, has been actively resisted by Republican party representatives - as well as by various employer bodies (Hendrick, 1997). Nevertheless, there are a number of different states that are in the process of trying to produce their own legislation in this area. For instance, California currently operates under a watered down version of the above OSHA standard, whilst in Washington State legislation is expected on safety and health consultation through employee committees.

In terms of guidance and standards, the Ergonomics Program Management Guidelines for Meatpacking Plants (OSHA, 1991) describe "major elements" which employers should incorporate in their own ergonomics programmes. The requirement for employee involvement is very clearly stated and recommendations are made for safety and health committees, ergonomic teams and employee suggestion procedures. In order to show how such an approach may be used, NIOSH produced a report describing in detail three intervention projects in the meatpacking industry (Gjessing et al 1994). In addition to describing the way in which the participation was structured and implemented, the report summarises what can be learned from these cases. Although these guidelines were produced for the meatpacking industry, they have been used to good effect elsewhere (Cohen, 1997). The draft ANSI standard Z-365 (Control of Work-Related Cumulative Trauma Disorders) describes employee participation as "essential to the success of the program" stating that employee participation should include, amongst other things, formal teams and meetings, informal discussions and employee surveys.

9.0 Case studies in participatory ergonomics

Throughout this report there have been many examples of 'lessons' taken from a whole variety of participatory ergonomics applications. In this section a table is presented (table 4) which summarises some of the information from these and a number of other case studies, with particular regard to the issues of health and safety. For each case referenced, in addition to stating the country and 'industry' in which the work took place, a brief description of the aims of the initiative, the participatory structure employed and any methods or techniques referred to are provided. Upon reflection our interpretations of the success of the initiative have deliberately been excluded, as it was felt that the complex issue of evaluating participatory ergonomics could not usefully be dealt with in this way (see section 7.0).

Authors	Country	'Industry'	Aims of initiative	Participatory structure	Participatory methods or techniques reported
Garmer et al, 1995	Sweden	vehicle production	development of ergonomics improvement process	work teams with rotating leadership function (already part of work organisation)	dialogue model (defines roles and competencies of operators and engineers), ergonomics knowledge package
Pransky et al, 1996	USA	paper products manufacturer	ongoing process to reduce musculoskeletal disorders	all employees members of work teams responsible for quality, productivity and ergonomics in their work area, team reps elected to plant ergonomics committee	problem solving process
Lewis et al, 1988	USA	large corporation	ongoing process to address productivity, quality and health and safety problems	various problem solving teams (as part of organisational participative management approach)	brainstorming, weighted voting, video
Jergerlehner, 1995	USA	industrial equipment manufacturer	ergonomics programme to reduce musculoskeletal disorders	ergonomics co-ordinator plus full-time ergonomics support team (in post for one year before rotating)	employee interview and suggestion form, checklist
Faville, 1995	USA	aircraft assembly	ergonomics programme to reduce musculoskeletal disorders	core group plus a multifunction team within each factory area	observation, checklists

Table 4: Case studies in participatory ergonomics or other relevant areas

Authors	**Country**	**'Industry'**	**Aims of initiative**	**Participatory structure**	**Participatory methods or techniques reported**
Keyserling and Hankins, 1994	USA	vehicle production	ergonomics programme to reduce musculoskeletal disorders	plant based ergonomics committees	checklists
Jones, 1997	USA	poultry processing	ergonomics programme to reduce musculoskeletal disorders	corporate programme structure, ergonomics committees	checklists, work site analysis
Moore and Garg, 1997	USA	meat processing	ergonomics programme to reduce musculoskeletal disorders and improve productivity	plant ergonomics team, departmental teams	forms and checklists
St Vincent et al, forthcoming a and b	Canada	electronics	ergonomics programme to reduce musculoskeletal disorders	ergonomics committee (operators and technical specialists) supplemented by workers from area under investigation	sampling plans, video, interviews, risk factor identification and ranking, mock-ups, simulations
Haims and Carayon, 1996	USA	office work	ongoing programme to identify and solve ergonomics problems, education of other workers	group of ten "ergonomics co-ordinators"	not reported

Table 4: Case studies in participatory ergonomics or other relevant areas (cont.)

Authors	Country	'Industry'	Aims of initiative	Participatory structure	Participatory methods or techniques reported
Mansfield and Armstrong, 1997	USA	library	ergonomics programme to control risks of musculoskeletal disorders and improve worker comfort and efficiency	co-ordinating committee, departmental ergonomics committees	interviews, checklists, surveys, video
Liker et al, 1991 (two cases)	USA	vehicle production	ergonomics programmes to reduce musculoskeletal disorder	steering committee plus various task forces and advisory groups	statistical analysis measurement (SAM) video, computer analysis, brainstorming
Gjessing et al, 1990 (three cases)	USA	meat processing	ergonomics programmes to reduce musculoskeletal disorders	(i) and (ii) plantwide ergonomics committee plus multifunction departmental teams (iii) multifunction teams (including top plant officials)	(i) brainstorming, problem analysis and prioritisation (ii) checklists and forms (iii) brainstorming, solution prioritisation
Wilson, 1995	UK	incinerator plant	workplace redesign (eventually developed into continuous improvement process)	problem solving group	adapted design decision group

Table 4: Case studies in participatory ergonomics or other relevant areas (cont.)

Authors	Country	'Industry'	Aims of initiative	Participatory structure	Participatory methods or techniques reported
Wilson, 1991a	UK	supermarket checkout design, library	workplace redesign	problem solving groups	design decision group
Kragt, 1992	Netherlands	chemicals	control room redesign (eventually developed into continuous improvement process)	input from all relevant engineers and shift workers	questionnaires, scale modelling, equipment testing
Kukkonen and Koskinen, 1993	Finland	office work	workplace redesign	all relevant staff	checklist
Lehtela and Kukkonen, 1991	Finland	telecommunications	workplace design	project group composed of worker representatives	design decision group
Snow et al, 1996	USA	University	workplace design	participatory design group (composed of stakeholder representatives)	scenario driven discussions and dynamic space manipulation using CAD drawings, virtual reality mock-up and walkthrough
Imada and Stawowy, 1996	USA	food service	workplace redesign	participatory design group (composed of stakeholder representatives)	design decision group type format

Table 4: Case studies in participatory ergonomics or other relevant areas (cont.)

Authors	Country	'Industry'	Aims of initiative	Participatory structure	Participatory methods or techniques reported
Moore, 1994	USA	vehicle production	improvement of an engineering task	multifunction group	brainstorming, prioritised voting
Orifice, 1996	USA	pharmaceuticals	reduction of physical demands associated with a particular operation	project team composed of engineers and operator representatives (assumed roles of customer, supervisor, quality assurance and human resources)	pre-approved funding, group problem solving, prototype development (in association with vendor)
Kuorinka et al, 1994	Canada	design of emergency service vehicles	workplace design project	six project groups (semi - experimental setting)	trace method (frequency analysis of equipment usage) drawings, mock-ups
Algera et al, 1990	Netherlands	steel industry	(i) workplace design project (ii) job and organisation design following the introduction of high-technology equipment	relevant operators and management	(i) interviews, video, discussions, scale modelling, mock-ups (ii) interviews, problem prioritisation, discussions
Buckle and Ray, 1991	UK	publishing (two cases)	introduction of new technology	user group (representatives of those directly involved with change)	user discussion groups to identify problems and identify needs and goals

Table 4: Case studies in participatory ergonomics or other relevant areas (cont.)

Authors	Country	'Industry'	Aims of initiative	Participatory structure	Participatory methods or techniques reported
Hornby and Clegg, 1992	UK	banking	development and implementation of a new system	various project groups to develop new system and manage organisational change	informal discussions, semi-structured interviews with user representatives, questionnaires
Zink, 1996	Germany	(i) electronics (ii) vehicle production	(i) introduction of new technology across an organisation (ii) as above but aimed to develop project into a continuous improvement process	project group assisted by small core teams (project required cross department co-operation)	(i) workshops (to undertake requirement analysis) (ii) discussion groups, problem solving activities
Frei et al, 1993	Switzerland	electronics	development of a new production process (work and organisation design plus design of workplace layout)	project steering committee and two project groups (composed of worker representatives)	planning meetings recorded using 'metaplan technique', layout modelling using construction sets
Brenner and Ostberg	Sweden	office work	design and planning of new office automation system	various project groups	not reported
Aikin et al, 1994	Australia	gas products manufacturer	incorporation of ergonomics into prospective design and development activities	all those responsible for the design of plant and equipment	ergonomics design review process (checklists plus other analytical tools)

Table 4: Case studies in participatory ergonomics or other relevant areas (cont.)

Authors	Country	'Industry'	Aims of initiative	Participatory structure	Participatory methods or techniques reported
Carter and Boquist	Sweden	construction	product design (mechanical aid for tile laying)	project group composed of representatives of all stakeholders	function identification and solution generation
Garrigou et al, 1995	France	newspaper printing	modernisation of plant (prospective design)	working group (technical and ergonomic function) 'follow-up group' (management and staff representatives - responsible for effective implementation)	activity analysis
Kuorinka and Patry, 1995	Canada	(i) steel industry (ii) poultry processing (iii) drinks distribution centre (iv) refrigerator manufacturer	(i) redesign of a production process (ii) reduction of musculoskeletal disorders (iii) reduction of lower back pain and accidents (iv) to address problems associated with a new production line	(i) multifunction group (ii) research team (group of 'experts') undertake analyses participatory group validate and implement recommendations (iii) group of supervisor and worker representatives (iv) group of workers and supervisors	(I) MTM analysis, simulation of solutions (ii) not reported (iii) work analysis (iv) not reported

Table 4: Case studies in participatory ergonomics or other relevant areas (cont.)

Authors	Country	'Industry'	Aims of initiative	Participatory structure	Participatory methods or techniques reported
Nagamachi, 1995; (three cases)	Japan	(i) air conditioning manufacturer (see also Nagamachi, 1991) (ii) vehicle production (iii) truck assembly	(i) redesign production line (ii) redesign jobs to improve job satisfaction (iii) accident reduction	steering committee / project team plus quality circles	(i) JDLC, Nagamachi's working posture assessment tool, 2D mannequin (ii) discussion and experiment (iii) not reported
Nagamachi, 1991	Japan	(i) car engine assembly (ii) agricultural machine manufacturer	(i) prevention of back injuries (ii) introduction of automatic production systems	(i) multifunction project team, group composed of assembly line managers, quality circles (ii) quality circles	(i) Nagamachi's working posture assessment (ii) discussion, use of models
Vink et al, 1995	Netherlands	office work	project to reduce operator workload	steering committee, project team, departmental members	observation, checklists, questionnaires
Westlander et al, 1995	Sweden	VDT workplaces	improvement of VDT work (short term intervention project initiated with aim of developing ongoing improvement process)	research team (expert group) working groups composed of departmental workers	workshops, proposal prioritisation
Laitinen et al, 1997	Finland	railway vehicle repair	improvement of "industrial housekeeping"	departmental teams (manager, supervisors, safety representative, workers)	checklists

Table 4: Case studies in participatory ergonomics or other relevant areas (cont.)

Authors	Country	'Industry'	Aims of initiative	Participatory structure	Participatory methods or techniques reported
Nagamachi and Tanaka, 1995	Japan	chemical fibre production	productivity improvement, development of jobs which are accessible to ageing workforce	project team (steering function) plus two multifunction sub teams	questionnaires, Job redesign for life cycle process (JDLC), video, working posture analysis
Moir and Bucholz, 1996	USA	construction	part of ongoing research programme into ergonomics and industrial hygiene interventions	various participatory groups (in order to span hierarchy in construction industry)	intervention ideas
Chaney, 1969	USA	computer manufacturer	semi-experimental setting - to investigate effect of employee participation in job design activities on performance and attitudes	supervisor-led employee groups	group goal setting and problem solving techniques

Table 4: Case studies in participatory ergonomics or other relevant areas (cont.)

10.0 Towards a conceptual framework for participatory ergonomics

The central thesis of this report is that participatory ergonomics is not a unitary concept. Participatory ergonomics initiatives can take a wide variety of forms, from a single redesign exercise to a full-blown culture of employee involvement. The aim of this section is to provide an insight into this diversity through identifying some of the dimensions across which participatory ergonomics initiatives might vary (see table 5). In doing so it draws on the work of, amongst others, Liker et al, 1989 Dachler and Wilpert, 1978 and Cohen, 1996, as well as our own previous fundamental research work (e.g. Wilson 1991b, Wilson and Haines, 1997). It should be noted that in describing these dimensions the perspective of Western industrialised countries is taken and, within that, a health and safety perspective. We are aware that under different circumstances these dimensions may well look quite different. For example, in south east Asia the only feasible route to ergonomics change might be participation across a community (see, for example, Kogi 1991; Kogi et al, 1988).

Dimension	Range
Extent / level	Organisation ______ Worksystem ______ Workplace ______ Product
Purpose	Work organisation ______ Design ______ Implementation
Continuity	Continuous ______ Discrete
Involvement	Direct (full/partial) ______ Representative
Formality	Formal ______ Informal
Requirement	Voluntary ______ Compulsory
Decision-making	Workers decide ______ Consensus ______ Consultation
Coupling	Direct ______ Remote

Table 5: Dimensions of participatory ergonomics

The first dimension of participatory ergonomics is its **extent / level**. This concerns where participatory ergonomics is applied, whether across an organisation, a worksystem or a single workstation or product.

The second dimension of participatory ergonomics concerns its **purpose**. Is it being used to implement a particular change or to be *the* method of work organisation (whether under conditions of change or not) ? Use of participative techniques within a design exercise is often part of implementing a change, although participatory techniques may also be used as part of a product design process.

A further dimension is provided by the **continuity** of use of participation. Has the process got a continuous or discrete timeline: is participation to be used as an everyday part of an organisation's activities or is it applied from time to time as a one-off exercise ?

The next dimension of participatory ergonomics - **involvement** - concerns who will actually take part in the process. On the one hand, there is full direct participation, when all stakeholders directly affected become participants (however, restrictions on resources or large numbers of potential contributors may dictate that they are represented by a sub-group of those affected - termed partial direct participation). On the other hand there is representative participation, which generally occurs in two situations; firstly where worker representatives are involved in product or equipment or job design, but may not necessarily be the eventual users and, secondly, where trade union personnel represent the interests of their members at work.

The fifth dimension acknowledges that the **formality** of worker participation in ergonomics may vary. In many cases involvement will be arranged through formal mechanisms such as teams or committees. In other cases worker participation will take place on an informal basis (Dachler and Wilpert, 1978).

The sixth dimension of participatory ergonomics concerns the **requirement** for participation - is it voluntary or compulsory ? Voluntary participation is the most usual form, in the sense that participation works best where the workforce volunteers its contributions and is involved in setting up the process. Compulsory participation is seen, for instance, in companies with compulsory quality circles or production groups, where involvement in troubleshooting and continuous improvement is an obligatory part of job specification and roles; it is arguable whether such compulsory participation is truly participative.

The seventh dimension considers **decision-making** structures. At one end of the scale, workers are consulted, although decisions rest with management. At the other end of the scale, decisions are made by workers. The final category concerns the situation where input is obtained from a range of those affected and decisions are made by consensus.

Finally, how directly participative methods are applied (a concept referred to as **coupling**) may vary. Participative methods may be directly coupled, where participants' views and recommendations are applied relatively directly (work groups redesigning their own workplace, for example). Remote coupling involves some filtering of participants' views, for example through the use of company-wide questionnaires.

Figure 2 shows a first general framework developed by the authors to illustrate the bases on which an organisation might initiate and structure participatory ergonomics initiatives. It starts at the point where an organisation makes the decision to employ participatory ergonomics and identifies some of the main factors which may motivate this decision. Some form of initiative will then be implemented, the structure of which may be defined across the eight dimensions of participatory ergonomics already described. Criteria influencing both the structure of the initiative and the selection of participatory methods have been identified. An evaluation of the initiative will contribute to any further motivation to employ participatory ergonomics. The whole process will take place within an organisational and environmental context which will influence all its elements. This first framework needs testing and further development within real settings, and this should be considered as one possible extension of this study.

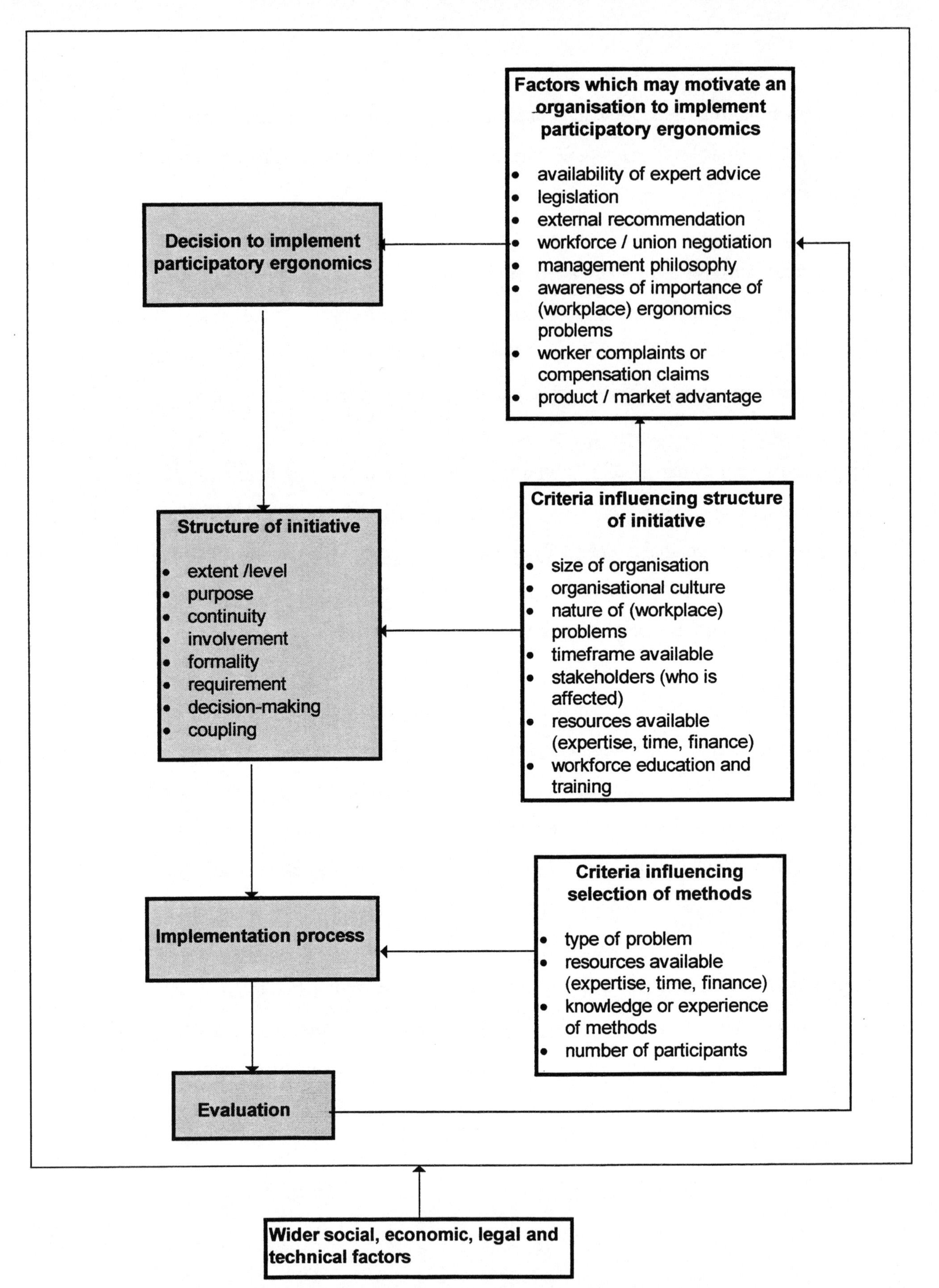

Figure 2: A first general framework illustrating the impetus and structure of participatory ergonomics initiatives

11.0 Requirements for participatory ergonomics

The participatory ergonomics literature shows some agreement about the requirements for a successful ergonomics initiative. In this section the aim is to review these requirements as well as looking at some of the recommendations from the change management and systems implementation literature. In doing this, it is important to remember that because participatory ergonomics can operate at many levels, from a single design exercise up to the basis of a shopfloor culture, some requirements will be more relevant than others depending on the type of initiative being undertaken.

11.1 Establishing a climate and support for participation

The organisational climate within which a participatory initiative is planned is important. It is difficult to implement any kind of change within a hostile or financially troubled environment and, therefore, participatory ergonomics should not generally be introduced in these circumstances. Cases have demonstrated that changes in the economic climate, job security or production processes not only affect practical issues such as the time and resources available to deal with ergonomics, but they are also likely to affect people's perceptions of the value or success of the project itself (see Marchington et al, 1994; Westlander et al, 1995; Buckle, 1996).

It is also important to bear in mind that there may be organisational attributes which particularly suit participatory ergonomics. These are likely to be open and less hierarchical structures, a history of good labour relations, a tradition of consultation, well established communication channels and job designs which emphasise personal control, responsibility and teamwork (Orifice, 1996 describes a successful participatory ergonomics project within the pharmaceutical industry in just such an environment). Of course, organisations which embody few of these key attributes may still benefit from participatory ergonomics, but it is likely that they will not do so to the same extent.

An important part of developing a climate for participatory ergonomics is establishing commitment for the programme throughout the organisation. Support from top management is, of course, essential and must be secured at the outset (e.g. see Liker et al, 1991; Gjessing et al, 1994). This means that the participatory process must be approved at Board level (or similar) and that this must be translated into active support when and where necessary. A rational cost-benefit case can often play an important role in obtaining this support. In a case from the poultry processing industry, Jones (1997) describes how meetings with the risk management department helped gain top management support for an ergonomics programme. It is also important that management make their commitment to the programme obvious. Cohen (1996) describes a number of ways in which this can be done. One approach is for management representatives to sit on steering committees and another is for a formal policy to be written that allows decisions about working conditions to be delegated to relevant participatory groups. A further demonstration of commitment, which is also an operational requirement, can be provided by allocating resources to implement changes recommended through participatory mechanisms.

The issue of resources is clearly very important. Any participatory process needs sufficient financial support both to set it up (where costs may include providing ergonomics training or external support from an ergonomist, e.g. see Mansfield and Armstrong, 1997) and to be able to actually implement any resulting changes. In a number of instances it has been found to be useful to set some kind of budget at the outset (Vink et al, 1995; Wilson, 1995b). Sufficient time will also need to be allocated. A participatory ergonomics process is likely to be fairly time consuming in comparison to a more traditional ergonomics approach (Buckle, 1996; Vink et al, 1995; Haims and Carayon, 1996) and this needs to be catered for. As well as appreciating the time requirements for the overall process, time must be made available for those job holders involved to meet and undertake their ‘participatory ergonomics duties’.

The importance of securing support from the top has already been mentioned, however, the need for support does not stop there. If middle management and supervisory staff are not supportive of participatory ergonomics it will be difficult to sustain any kind of on-going process. Cohen (1996) reports that there is some suspicion that the failure of many of the quality circle initiatives started in US companies was at least partly due to resistance from middle management, for example due to the fear of loss of power or status. Support must also be sought amongst both the job holders and their trade unions. Many participative changes will have implications for the terms and conditions of employment and may be subject to collective bargaining. Some resistance is to be expected - and it is important to remember that resistance to change is not always irrational (see section 5.0). In these situations open discussion of the causes of any resistance can help considerably. Whilst the need to secure support from throughout the organisation may seem obvious, knowing exactly when to inform the people affected of an impending participatory ergonomics initiative is nothing like so clear cut. Inform them too late and rumours may have spread and opposition hardened; tell them too early and if, as often happens, delays arise, then the process is likely to fall into disrepute as people wait for something to occur. Finally, it should be understood that it is difficult for people to give their support unless realistic, flexible goals have been set and any constraints made clear to everyone. For instance, Vink et al (1995) stress that the project goals and the framework by which it will proceed should be agreed at the outset by all those involved, meaning that decisions made further down the line are also more likely to be accepted (even if for some they are likely to have unfavourable consequences).

11.2 Structuring a participatory initiative

The setting-up of any given participatory initiative brings with it a further set of important considerations. The first of these concerns the form or structure by which the programme will progress. Within any on-going process, it seems important to create something sufficiently structured and yet still flexible enough to allow it to respond as the process and participants develop. Flexibility is also important to allow the programme to adapt to any changes in the wider organisational context (Haims and Carayon, 1996). There are good arguments both for starting a project on a small scale and then allowing it to 'grow' - the idea being that the use of 'micro-level' participative processes will prepare the ground for more extensive participative strategies to be put in place - as well as for setting up organisation wide structures to support participative practices from the outset (Wilson and Haines, 1997; Joseph, 1997). Mansfield and Armstrong (1997) argue that gradually implementing a participatory programme in 'pieces' provides time for people to take on board new ideas and allows parts of the programme to be evaluated and changed if necessary.

A number of authors have mentioned the benefits of having a steering committee oversee company-wide or continuing ergonomics programmes (Vink and Urlings, 1995; Gjessing et al, 1994). According to Liker et al (1991) a joint union-management steering committee provided the necessary leadership to give their ergonomics programme direction (particularly in its initial stages). It was also thought to play an important role in motivating middle managers to support the programme. Although the exact role of a steering committee may vary somewhat, it is largely there to perform an advisory function and shouldn't attempt to take over the more 'micro level' aspects of the programme. Liker et al report how a steering committee's attempt to redesign jobs lead to recommendations for change which proved unfeasible.

It is vital to have representatives of all the main stakeholders within the participatory process. Cohen (1996) describes how the analysis of ergonomics problems and the implementation of improvements typically requires a multidisciplinary response (see also Imada, 1991). In addition to workers and management, he lists engineering, maintenance, safety, occupational health, human resources and ergonomics amongst the other groups who may usefully contribute to an ergonomics initiative. Two cases have highlighted the importance of involving representatives from those groups who actually implement changes. In a case from the automobile industry, many of the ergonomics improvements involved changes being made to existing plant machinery and equipment (Liker et al, 1991). The authors argue that this process was facilitated by the presence of skilled trade workers and the head of maintenance on the ergonomics committee. In contrast, the authors of a second case argue that the fact that the purchasing department was not represented on the steering committee of a participatory programme meant that it took a long time for new office furniture to be provided (Vink et al, 1995).

Worker input into the participatory process is, of course, essential. This will need to be structured in such a way that those with knowledge and experience of the workplace problems are involved. How this might be achieved can, of course, vary widely. In some cases, for instance, where the workforce of one production area analyse and redesign their own work environment, all of the group may participate (although some may not wish to be involved and this must be respected). In other cases, the participative group will be formed by calling for volunteers, holding elections or selecting participants. In still other instances, a mixture of these strategies may be employed. All strategies hold dangers of alienating some of the workforce or of disruptive influences in the group, and so care must be taken. Regardless of how the group is formed, it is important that, from the outset, participants are made aware of any limits upon what they can either know about or do.

There are two further issues regarding participant representation which should be mentioned at this point. Firstly, participatory processes may lead to participants knowing 'too much'. If people participate in a series of initiatives or else over a long period of time, they may cease to be representative of a 'typical worker', and start to view all issues from a design, engineering or ergonomics viewpoint. Of course, if this is widespread it can be a substantial gain for the company itself (see section 11.3) but the downside is that such participants may lose the viewpoint and insight which made their early contributions so valuable. Secondly, there is the potential difficulty that employees centrally involved in the participative process may become alienated from the colleagues they represent. In this situation, rather than continuing to be seen as one of the workers they are considered to have "gone over to the other side". An early case of this was described by worker directors at British Steel (Bank, 1977).

11.3 Participatory processes and methods

Many authors have highlighted the fact that those involved in participatory ergonomics programmes need appropriate training. Sometimes training in general ergonomics principles will be necessary; this must be pitched correctly as too much training is costly and time-consuming, whilst too little runs the risk of ergonomics being 'trivialised' - see St Vincent et al (forthcoming a). Participants may need to be coached in the specific ergonomics skills that they will use within the participatory process. It is also recommended that members of ergonomics committees or problem solving groups are given training in teamwork and interaction skills to allow them to perform effectively as a group (Gjessing et al, 1994). According to Liker et al (1991: pp. 134) "ergonomics committees that are designed and trained to be participative and develop an effective group process will keep members motivated and do better quality job redesign." Finally, it is not only the workers who are likely to benefit from training, management, too, may have to acquire new skills. For example, Gjessing et al (1994) state that management may need instruction in how to relate to workers who have taken over decision-making roles. Furthermore, in any ongoing participatory ergonomics process, learning requirements are likely to change over time as both the participants and the process develops. For instance, Haims and Carayon (1996) describe how initial needs for training in ergonomics skills gave way to training in change management skills. As a final point, it is interesting to note that "train-the trainer" programmes have been increasingly used in ergonomics programme management as well as for health and safety generally (see Corlett, 1991; Silverstein et al, 1991).

Most participatory ergonomics initiatives will involve input from an ergonomist. This input can vary widely in terms of its scope and 'intensity'; sometimes the expert might be required to keep a fairly constant eye on proceedings whereas, on other occasions, their involvement may be more sporadic. One could imagine, for example, that the ergonomics consultant might play a particularly active role at the beginning, either in helping to set up the participatory process or in training the participants to analyse jobs or workplaces. Then, once this initial period is completed, the ergonomist can take more of a 'back seat' position to allow the participants to assume more responsibility for themselves (St Vincent et al, forthcoming b; see also Haims and Carayon, 1996 and Westlander et al, 1995 for discussions of this so-called 'internalisation' of participatory ergonomics programmes). Alternatively, the participants may be encouraged to get on with seeking solutions to a particular ergonomics problem, with the expert re-entering the scene near the end in order to check upon the overall validity and/or feasibility of their suggestions (see Vink et al, 1992; Wilson, 1995b).

The development of in-house expertise is described in some detail by Wilson (1994). Here Wilson argues that, whilst the challenge of 'Ergonomics Devolution' is to motivate people throughout a company or organisation to become their own ergonomists and to provide them with the necessary tools, techniques and training, they must also learn to recognise where and when specialist assistance is needed (this last point is reiterated by a number of other authors including St Vincent et al, forthcoming b; and Liker et al, 1991). For instance, by analogy, a good general ergonomist might recommend appropriate light levels for a certain task but will call upon a lighting engineer to specify the detailed requirements of the luminaires. Participants must have both the knowledge and confidence to understand the bounds to what they can do and also need the authority to call upon professional expertise (internal or external) where necessary. Such authority and the ready support of relevant expertise are especially important if participatory ergonomics is used as a basis for interventions to provide legal compliance or reduce potential for compensation claims.

Participation, by its very nature, will involve group or team-based activities. Recommendations for team size vary a little (for example, 6-12 or 7-15) but it is generally agreed that group interaction will be impaired if the team becomes too large. One way of incorporating larger numbers of participants is to set up a series of sub-groups or parallel groups. Moreover, many group processes, irrespective of whether they are part of an ongoing initiative or taking place on a 'one-off' basis, will involve a facilitator. This role is undoubtedly a difficult one, necessitating as it does a broad range of abilities. Ideally a facilitator should be willing to listen to others and yet capable of making firm decisions.

They should be seen as knowledgeable without being perceived as dogmatic. What is more, they should be able to recognise the point at which, having served their useful purpose, it is time to leave the process (see Buchanan and Boddy, 1992; Wilson, 1995b).

One topic of interest, frequently raised but with little guidance or empirical evidence, is the preferred composition of participant groups. A major question is whether the process is assisted or hindered by the presence of technical or organisational experts from within the company (e.g. manufacturing engineers or managers). The trade-off is of the additional expertise available versus the danger that others will be either cowed into not participating or else will use the opportunity for excessive confrontation.

In the definition of participatory ergonomics given at the beginning of this report, the need for participants to be motivated and knowledgeable was highlighted. However, there is another key requirement which inter-relates with both of these; namely confidence. As participants gain in knowledge and ability, they will slowly gain confidence both in their own contributions and also in the participative process itself - as they see for themselves that they are actually influencing events and outcomes. This, in turn, should motivate them and others to become more involved in the participative initiative. This close association between knowledge/ability, confidence and motivation underpins much modern theory of motivation at work (e.g. Expectancy Theory - Vroom, 1964). A model of this process is presented in figure 3. Early success appears vital. For this reason, Liker et al (1991) recommend that groups tackle fairly straightforward problems before attempting anything too complex (see also Gjessing et al, 1994).

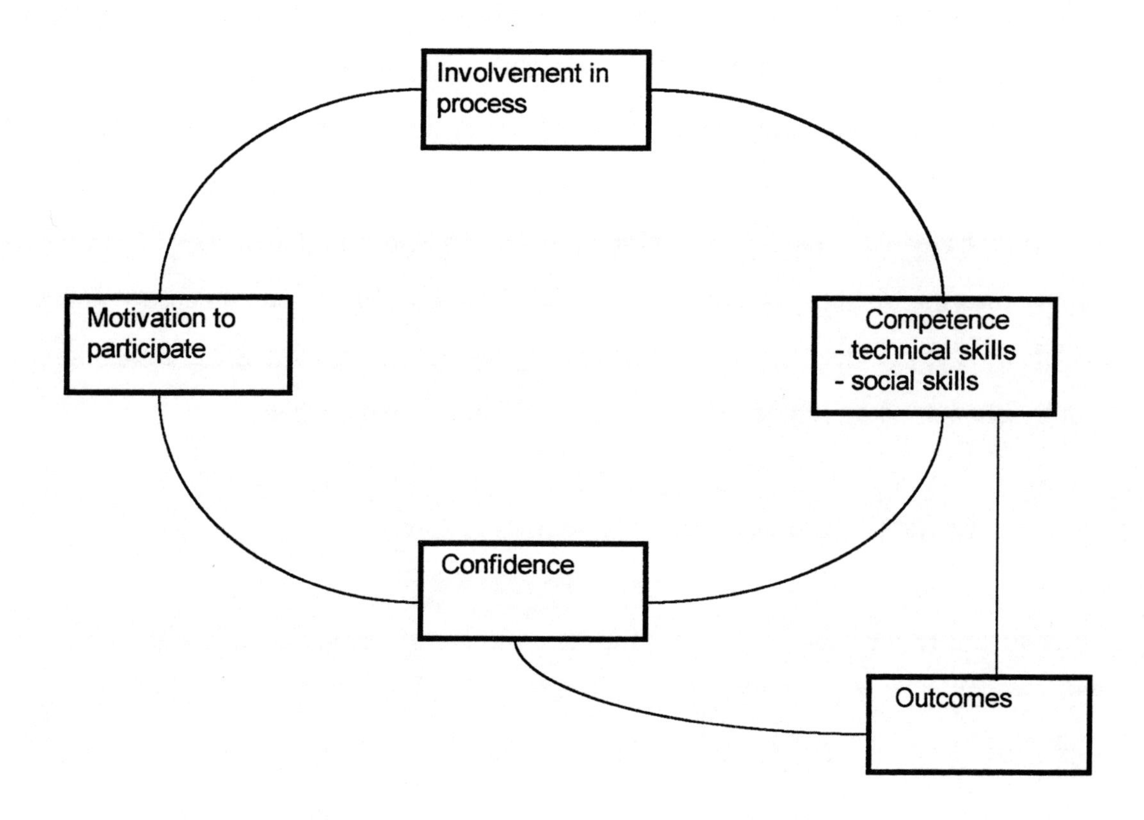

Figure 3: The participatory ergonomics cycle

As the participatory programme progresses there may be times when participants' motivation needs to be bolstered or reinforced. For example, workers may have developed unrealistic expectations about the time it takes to obtain management approval for their recommendations. They might also become frustrated at delays in both implementing recommendations and seeing the benefits (Gjessing et al, 1994). Difficulties may also arise as participants move from relatively simple problems to those of a more intractable nature.

On the subject of methods, it is important to recognise that how one goes about a participatory initiative will vary depending upon the particular remit. However, as a general rule, methods should be both cost-effective and as flexible as possible. The whole process should be iterative with sufficient room for compromise. The best methods tend to be those that promote creativity, support learning and inspire confidence. Goal setting and the provision of feedback information are just two devices that some authors feel can be useful (e.g. Haims and Carayon, 1996; Liker et al, 1991). In any case, participatory groups require sufficient information to allow them to carry out their tasks effectively. Therefore, management may have to make information available to participants that they might not normally have. For example, prioritising problem areas will require access to accident or injury records, whereas justifying recommendations will require data on costs and benefits (Cohen, 1996). Indeed, Mansfield and Armstrong (1997) see 'equalising' access to information as a key component of many participatory programmes.

11.4 Evaluating participatory ergonomics initiatives

It is obviously important to evaluate the effectiveness of any given participatory programme for not only does it enable adjustments to be made to ongoing initiatives but it also provides information which can be invaluable for future initiatives. This evaluation should not just consider the outcomes of the project, but it should also account for stakeholders' experiences of and responses to the participatory process as well as its efficiency and timeliness. As discussed earlier (see section 7.0) these evaluations can take a number of forms including process measures and cost-benefit calculations. As with job design change perhaps the best evaluation measure, although not explicit or financially based, is the satisfaction of the participants and the extent to which they and others elsewhere in the organisation wish to continue or to repeat the exercise.

12.0 Research requirements

One of the key points to have come out of this examination of the literature is that it is impossible to argue that any one form of participatory ergonomics or worker participation is appropriate in all circumstances. Considering the differences between organisations in terms of their cultures, their resources and the environments in which they operate and between the problems they face, it is not surprising that it is not possible to specify a single 'best' way. Indeed, it could also be argued that any assumption that there is a single 'best way' to achieve participatory ergonomics goes against the very spirit of participation. Recognition of this necessary diversity in approach should not, however, preclude research into participatory ergonomics. As Haims and Carayon (1996) point out, the implementation of participatory ergonomics programmes has largely been a process of "trial and error". They point to the lack of any theoretical framework and argue that what is needed are principles to guide the implementation of participatory ergonomics programmes within organisations (see also Haines and Wilson, 1997).

Overall, the participatory ergonomics literature is largely characterised by case studies. Of course, there is much to be learned from these cases - as much from the failures as the successes (Westlànder et al, 1995). Some commentators, such as Cohen (1996) and Vink et al (1992) have urged that more initiatives across a wider field need to be reported, especially as many of the existing cases come from larger organisations and often in manufacturing. For although there appears to be some evidence of consistency across different case studies in terms of the factors which go together to form a successful participatory ergonomics initiative, these have not been studied in anything like a systematic fashion (Cohen, 1996). This is certainly an area where further work is needed.

To take one important area, one of the most commonly reported requirements for participatory ergonomics is the need for management commitment. In gaining this commitment consideration of the value of participatory ergonomics and the resources required are very important. One of the arguments often given against taking a participative approach to ergonomics is that it is very time consuming, on the other hand many of the proponents of participation counter this by suggesting that there are savings in time and/or resources which are made later on in the process that more than offset the initial additional costs. Even so, there do not appear to be any studies reported which compare cost benefit analyses of participatory ergonomics initiatives to more traditional approaches to ergonomics, possibly due to the potential expense and difficulty in undertaking such work.

A further example of an area where there has been little systematic examination concerns some of the structural elements of participatory ergonomics. Here issues of interest might include the preferred composition of participatory groups (for example, does the presence of engineers or management help or hinder the participatory process for shopfloor groups ?) or a comparative evaluation of different participatory tools or methods.

It is interesting to note that much of the existing literature on participatory ergonomics stems from outside of the UK, and it is unclear to what extent work is being carried out closer to home. Is a considerable body of work being undertaken but, for a variety of reasons is largely unreported ? It would seem useful, therefore, to survey UK companies to identify where ergonomics programmes are already in place and to find out whether they incorporate some level of employee participation.

Overall, the main gap in research seems to be the lack of systematic studies of participatory ergonomics programmes. There are a number of possible reasons why this may be so, including the difficulty of setting up adequate controls and the expense of tracking large numbers of studies. Of course this is not a problem exclusive to participatory ergonomics, for this lack of 'scientific' intervention research characterises many other kinds of ergonomics projects too (see Buckle and Li, 1996). Another gap concerns the lack of published advice or guidance for UK companies wanting to implement a participatory approach to ergonomics. There is general agreement in the literature that specifying one 'off the shelf' method of setting up and running a participatory ergonomics programme would not be very useful. And although at present we can describe some factors that are reported as being important in particular successful initiatives, it may be more worthwhile trying to provide more specific guidance by describing a range of approaches (structures, methods) for participatory ergonomics. Companies could then select and adapt those approaches which would be most appropriate for their own circumstances.

13.0 Conclusions

There can be no denying that participatory ergonomics is currently the focus of a great deal of interest. It would appear that a rapidly expanding number of organisations are seeing participation as an important way of addressing work-related problems, including, significantly, those of health and safety. However, this developing enthusiasm for participatory ergonomics belies the fact that the concept is less than straight forward. As soon as one attempts to pin down what participatory ergonomics means, or what it consists of, the picture begins to blur. One possible explanation for the 'cloudiness' of the concept is that it is used inconsistently. That is, people are calling things "participatory" that, in reality, are not. However, it is more likely that the confusion is due to the fact that participatory ergonomics is a complex and diverse concept. It is, in other words, a kind of umbrella term used to cover a fairly broad range of ideas and practices. This conclusion helps to make sense of the fact that there is not one but a whole host of different definitions, not one but a range of models and ways of doing participatory ergonomics. It also makes sense of the fact that there are a multiplicity of tools and methods employed within participatory ergonomics initiatives.

Yet despite these variations, it would appear that most commentators see participatory ergonomics as offering a common set of advantages. Firstly, it is seen as exploiting the detailed knowledge and experience of those who inhabit the very workplace under investigation - getting most from those who, in a sense, should know best. Secondly, it is felt to encourage a sense of ownership amongst the participants, such that it helps to secure at least some degree of commitment both to the process itself and to any changes that may result. A third and related advantage of participatory ergonomics is thought to centre upon psychosocial factors. Irrespective of the specific details of how these effects operate, it is not surprising that a workforce whose views are sought and taken seriously might feel more positively about both themselves and their workplace (a fact which would seem to prioritise the more inclusive forms of participation). Against such benefits, there may be some concerns about the time, cost and ease of implementing participatory programmes, and about inappropriate or poor quality implementations. However, it is true to say that, on balance, many feel that the pros outweigh the cons.

If one accepts, therefore, that participatory ergonomics (however one defines it) has great potential, then questions arise concerning how one might effectively harness this. One of the aims of this report has been to take a step along this road by providing the outlines of a conceptual framework for understanding how participatory programmes may be initiated and executed, describing a number of different dimensions along which any given participatory project may be 'plotted'. It must be stressed, however, that this framework is speculative and provisional. What is now required is a much more detailed and systematic study of how participatory programmes function.

References

Aikin, C., Rollings, M., and Wilson, J.R., 1994, Providing a foundation for ergonomics: systematic ergonomics in engineering design (SEED). In *Proceedings of the Twelfth Congress of the International Ergonomics Association*, 5, 216-218.

Algera, J.A., Reitsma, W.D., Scholtens, S., Vrins, A.A.C., Wijnen, C.J.D., 1990, Ingredients of ergonomic intervention: how to get ergonomics applied, *Ergonomics*, 33, 5, 557-578.

Bank, J., 1977, *Worker directors speak.* Gower Press, Farnborough, Hants

Batt, R. and Appelbaum E., 1995, Worker Participation in Diverse Settings: Does the Form Affect the Outcome, and If So, Who Benefits? *British Journal of Industrial Relations*, 33(3).

Bernoux, P., 1994, Participation: A review of the literature. *European Participation Monitor.*

Bongers, P.M., and Houtman I.L.D., 1996, Psychosocial Factors and Musculoskeletal Symptoms; First Results of the Dutch Longitudinal Study. In Proceedings of the 25th International Conference on Occupational Health, Stockholm, Sweden Sept. 15-20 1996

Brenner, S., and Ostberg, O., 1995, Working conditions and environment after a participative office automation project. *International Journal of Industrial Ergonomics*, 15 379-387

Brown, O. Jr., 1990, Marketing participatory ergonomics: current trends and methods to enhance organizational effectiveness. *Ergonomics*, 33, 5, 601-604.

BS 8800: *1996 Guide to Occupational Health and Safety Management Systems.* British Standards Institution

Buchanan, D. and Boddy, D., 1992, *The Expertise of the Change Agent.* Prentice-Hall, London.

Buckle, P.W. and Ray, S., 1991, User Design and Office Workers - An Evaluation of Approaches. In: *Contempory Ergonomics, Proceedings of the Ergonomics Society's 1991 Annual Conference*, Southampton, England

Buckle, P. and Li, G., 1996, User needs in exposure assessment for musculoskeletal risk assessment In: *Virtual Proceedings of CybErg*, The First International Cyberspace Conference on Ergonomics (eds.) Straker L. and Pollock, C.

Buckle, P., 1996, Participatory Ergonomics and Risk Management. In *proceedings of the symposium Risk Assessment for Musculoskeletal Disorders*, National Institute of Occupational Health, Copenhagen, 13-14 September

Burgoyne, J.G., 1994, Stakeholder analysis. In *Qualitative Methods in Organizational Research: A Practical Guide*, C. Cassell and G. Symon (eds.) SAGE Publications, New York.

Caccamise, D.J., 1995, Implementation of a team approach to nuclear criticality safety: The use of participatory methods in macroergonomics, *International Journal of Industrial Ergonomics*, 15, 397-409.

Caplan, S. 1990, Using focus group methodology for ergonomic design, *Ergonomics*, 33, 527-533.

Carter, N., and Boquist, B., 1995, Participatory Ergonomics and the Development of a Mechanical Aid for Laying Concrete Paving Tiles In *Advances in Industrial Ergonomics and Safety VII*, Bittner, A.C. and Champney, P.C.,(eds) Taylor and Francis, London 437-441

Chaney, F.B., 1969, Employee Participation in Manufacturing Job Design, *Human Factors*, 11(2), 101-106.

Cherns, A., 1976, The principles of sociotechnical design. *Human Relations*, 29(8) 783-792

Cherns, A., 1987, Principles of sociotechnical design revisited, *Human Relations*, 40, 153-162.

Childs, J., 1984, *Organisation: A guide to problems and practice*, Harper & Row, London.

Cobb, S.V., D'Cruz, M.D., and Wilson, J.R., 1995, Integrated Manufacture: a role for virtual reality ? *International Journal of Industrial Ergonomics*, 16 411-425

Coch, L. and French, J., 1948, Overcoming Resistance to Change. *Human Relations*, pp.512-532.

Cohen, A.L., 1996, Worker Participation, In *Occupational Ergonomics Theory and Applications*, Bhattacharya, A. and McGlothlin, J.D., (eds), Marcel Dekker, New York 235-258.

Cohen, R., 1997, Ergonomics Program Development: Prevention in the Workplace, *American Industrial Hygiene Association Journal*, 58, 145-149.

Cohen, A., 1997, Personal Communication

Corlett, E.N., 1988, Cost-benefit Analysis of Ergonomic and Work Design Changes. In *International Reviews of Ergonomics*, D.J. Oborne (ed.) 2, 85-103. Taylor & Francis, London.

Corlett, E.N., 1991, Ergonomics Fieldwork: An Action Programme and Some Methods. In *Towards Human Work: Solutions to Problems in Occupational Health and Safety*, M. Kumashiro and E.D. Megaw (eds.). Taylor & Francis, London.

Cotton, J.L., Vollrath, D.A., Froggat, K.L., Lengnick-Hall, M.L., and Jennings, K.R., 1988, Employee participation: Diverse forms and different outcomes. *Academy of Management Review*, 13, 8-22

Cotton, J.L., 1993, *Employee Involvement. Methods for Improving Performance and Work Attitudes*. Sage, Newbury Park

Dachler, H.P., and Wilpert, B., 1978, Conceptual dimensions and boundaries of participation in organizations: A critical evaluation. *Administrative Science Quarterly*, 23, 1-39

Daniellou, F., Kerguelen, A., Garrigou, A. and Laville, A. 1990. Taking future activity into account at the design stage: participative design in the printing industry. In *Work Design in Practice* C.M. Haslegrave, J.R. Wilson and E.N. Corlett (eds.) Taylor & Francis, London.

Daniellou, F. and Garrigou, A., 1992. Human factors in design: Sociotechnics or ergonomics? In *Design for Manufacturability and Process Planning*, M. Helander and M. Nagamachi (eds.) Taylor & Francis, London. pp.55-63.

Diani, M. & Bagnara, S., 1984, Unexpected consequences of participative methods in the development of information systems: The case of office automation. In *Ergonomics and Health in Modern Offices*, Grandjean, E. (ed.) pp.227-232. Grandjean, E. (ed.) Taylor & Francis, London.

Eason, K., 1989, New systems implementation. In *Evaluation of Human Work: A Practical Ergonomics Methodology*, Wilson, J.R. & Corlett, E.N. (eds.) Taylor & Francis, London.

Eklund, J., Forthcoming, Ergonomics, Quality and Continuous Improvement - Conceptual and Empirical Relationships in an Industrial Context. To appear in *Ergonomics.*

Eklund, J., 1997, Personal communication

Faville, B.A., 1995, One Approach for an Ergonomics Program in a Large Manufacturing Environment, *Advances in Industrial Ergonomics and Safety VII*, 302-3

Forrester, K., 1986, Involving workers: participatory ergonomics and the trade unions. In *Contemporary Ergonomics*, D.J. Oborne (ed). Taylor & Francis, London.

Fraser, C., and Foster D., 1982, Social groups, nonsense groups and group polarization in Tajfel, H. (ed) The Social Dimension, Cambridge, Cambridge University Press

Frei, F., Hugentobler, M., Schurman, S., Duell, W. and Alioth, A., 1993, Alcatal STR: Bottom-up Approach. In *Work Design for the Competent Organization* Greenwood Press. Westport.

Garmer, K., Dahlman, S. and Sperling, L., 1995, Ergonomic development work: Co-education as a support for user participation at a car assembly plant. A case study, *Applied Ergonomics*, 26(6), 417-423.

Garrigou, A., Daniellou, F. Carballeda, G. and Ruaud, S., 1995, Activity analysis in participatory design and analysis of participatory design activity, *International Journal of Industrial Ergonomics, Special Issue: 'Participatory Ergonomics*, 15, 5, 311-329.

Gjessing CC, Schoenborn, TF and Cohen, A., 1994, Participatory Ergonomics Interventions in Meatpacking Plants, *DHHS (NIOSH)* Publication No. 94-124

Gjessing, C.J., 1997, Personal Communication

Glew, D.J., O'Leary-Kelly, A.M., Griffin, R.W. and Van Fleet, D.D., 1995, Participation in Organizations: A Preview of the Issues and Proposed Framework for Future Analysis, *Journal of Management*, 21(33), 395-421.

Griffith, R.W., 1985, Moderation of the effects of job enrichment by participation: A longitudinal field experiment, *Organizational Behaviour and Human Decision Processes*. 35, 73-93.

Grunberg, L., Moore, S. and Greenberg, E., 1996, The Relationship of Employee Ownership and Participation to Workplace Safety, *Economic and Industrial Democracy*, 17, 221-241.

Haims, M.C., and Carayon, P., 1996, Implementation of an 'in-house' participatory ergonomics program: A case study in a public service organization, *Human Factors in Organizational Design and Management*, Brown, Jnr., V.O. and Hendrick, H.W. (eds.) Elsevier Science B.V., 175-180.

Haines, H.M., and Wilson, J.R., 1997, Towards a framework for participatory ergonomics. In *Proceedings of the Thirteenth Congress of the International Ergonomics Association*, Tampere, Finland.

Health and Safety Executive, 1993, *Successful health and safety management:* HS(G)65, HMSO

Health and Safety Executive, 1996, *A Guide to the Health and Safety (Consultation with Employees) Regulations*, HSE Books

Hendrick, H.W., 1996, Good Ergonomics Is Good Economics. In *Proceedings of the Human Factors and Ergonomics Society 40th Annual Meeting*

Hendrick, H.W., 1997, Personal communication

Hornby, P. and Clegg, C. 1992, User participation in context: a case study in a UK bank, *Behaviour & Information Technology*, 11, 5, 293-307.

Imada, A.S. & Robertson, M.M., 1987, Cultural perspectives in participatory ergonomics. *Proceedings of the Human Factors Society 31st Annual Meeting*. 1018-1022.

Imada, A.S., 1991, The rationale and tools of participatory ergonomics. In *Participatory Ergonomics*, Noro, K. and Imada, A.S. (eds.) pp. 30-51. Taylor & Francis, London

Imada, A.S., 1994, Self-Learning Organizations and Participatory Strategies, *Proceedings of the Twelfth Congress of the International Ergonomics Association*, 6(1), 98-100.

Imada, A.S. and Stawowy, G., 1996, The effects of a participatory ergonomics redesign of food service stands on speed of service in a professional baseball stadium, *Human Factors in Organizational Design and Management*, Elsevier Science 203-208.

Jegerlehner, P.E., 1995, Workers' Participation Helps Reduce Cumulative Trauma Disorder Injuries, In *Advances in Industrial Ergonomics and Safety VII*, Bittner, A.C. and Champney, P.C., Taylor and Francis, London, 339-342

Jensen, P-L., 1994, Participatory Ergonomics Present Status and Central Issues seen from Scandinavia, *Proceedings of the Twelfth Congress of the International Ergonomics Association*, 6(1), 120-122.

Jensen, P.L., 1996, State Regulation and Development of Human Factor Activities in Firms, In *Human Factors in Organizational Design and Management*, Brown, Jnr., V.O. and Hendrick, H.W. (eds.) Elsevier Science 167-173

Jensen, P-L, 1997, Personal communication

Jones, R.J., 1997, Corporate Ergonomics Program of a Large Poultry Processor, *AIHA Journal*, 58, 132-137.

Joseph, B., 1997, Personal communication

Karasek, R. and Theorell, T., 1990, *Healthy Work: Stress, Productivity and the Reconstruction of Working Life*. Basic Books.

Karwowski, W., and Salvendy, G. (eds) 1994, *Organization and Management of Advanced Manufacturing*. John Wiley & Sons, New York

Keyserling, W.M. and Hankins, S.E., 1994, Effectiveness of Plant-Based Committees in Recognizing and Controlling Ergonomic Risk Factors Associated with Musculoskeletal Problems in the Automotive Industry, *Rehabilitation*, 3, 346-348.

Kirwan, B. and Ainsworth, L.K., 1992, A Guide to Task Analysis. Taylor & Francis, London.

Kogi, K., Phoon, W-O. and Thurman, J.E., 1988, *Low-Cost Ways of Improving Working Conditions: 100 Examples from Asia.* International Labour Office, Geneva.

Kogi, K., 1991, Participatory training for low-cost improvements in small enterprises in developing countries. In *Participatory Ergonomics*, Noro, K. and Imada, A.S. (eds.) pp. 73-81. Taylor & Francis, London.

Kukkonen, R. and Koskinen, P. 1993, User participation in workplace design, In *Work with Display Units 92*, Luczak H, Cakir A and Cakir G (eds). Elsevier.

Kuorinka, I., Cote, M-M, Baril, R., Geoffrion, R., Giguere, D., Dalzall, M-A, Larue, C. 1994, Participation in workplace design with refernce to low back pain: a case for improvement of the police patrol car. *Ergonomics*, 37, 1131-1136

Kuorinka, I. and Patry, L. 1995, Participation as a means of promoting occupational health, *International Journal of Industrial Ergonomics*, 15, 365-370

Kuorinka, I., 1997, Tools and means of implementing participatory ergonomics, *International Journal of Industrial Ergonomics*, 19, 267-270.

Laitinen, H., Saari, J., Kuusela, J., 1997, Initiating an innovative change process for improved working conditions and ergonomics with participation and performance feedback: A case study in an engineering workshop. *International Journal of Industrial Ergonomics*, 19 299-305

Leana, C.R., Locke, E.A., and Schweiger, D.M., 1990, Fact and fiction in analyzing research on participative decision making: A critique of Cotton, Vollrath, Froggatt, Langnick-Hall and Jennings, *Academy of Management Review*, 15 137-146

Lehtela, J. and Kukkonen, R. 1991, Participation in the purchase of a telephone exchange - a case study. Designing for Everyone, *Proceedings of the Eleventh Congress of the International Ergonomics Association*, Y. Queinnec and F. Daniellou (eds) Taylor and Francis, London.

Lewis, H.B., Imada, A.S., Robertson, M.M. 1988, Xerox Leadership through Quality: Merging Human factors and Safety through Employee Participation. *Proceedings of Human factors Society 32nd Annual Meeting*

Liker, J.K., Nagamachi, M. and Lifshitz, Y.R., 1989, A comparative analysis of participatory ergonomics programs in US and Japan manufacturing plants. *International Journal of Industrial Ergonomics*, 3, 185-189.

Liker, J.K., Joseph, B.S. and Ulin, S.S., 1991, Participatory ergonomics in two US automotive plants. In *Participatory Ergonomics*, Noro, K. and Imada, A.S. (eds.). pp. 97-139. Tayor & Francis, London.

Linstone, H.A. and Turoff, M. 1975, *The Delphi Method: Techniques and Applications*. Addison-Wesley, Reading, MA.

McClelland, I., 1995, Product assessment and user trials. In *Evaluation of Human Work*, Wilson, J.R. & Corlett, E.N. (eds.). Taylor & Francis, London.

McNeese, M.D., Zaff, B.S., Citera, M., Brown, C.E. and Whitaker, R., 1995, AKADAM: Eliciting user knowledge to support participatory ergonomics. *International Journal of Industrial Ergonomics, Special Issue: 'Participatory Ergonomics*, 15, 5, 345-365.

Mambrey, P., Opperman, R. & Tepper, A., 1987, Experiences in participative systems design. In *System Design for Human Development and Productivity: Participation and Beyond*, Doherty, P. et al (eds.) pp.345-358. North-Holland, Amsterdam.

Mansfield, J.A. and Armstrong, T.J., 1997, Library of Congress Workplace Ergonomics Program, *AIHA Journal*, 58, 138-144.

Marchington, M., Wilkinson, A., Ackers P. and Goodman J., 1994, Understanding the Meaning of Participation: Views from the Workplace, *Human Relations*, 47(8), 867-893.

Meister, D., 1985, *Behavioural Analysis and Measurement Methods*, J. Wiley & Sons, New York.

Moir, S. and Buchholz, B., 1996, Emerging Participatory Approaches to Ergonomic Interventions in the Construction industry, *American Journal of Industrial Medicine*, 29, 425-430.

Moore, J.S. 1994, Flywheel Truing - A Case Study of an Ergonomic Intervention. *American Industrial Hygeine Association Journal*, 55(3) pp 236-244.

Moore, J.S. and Garg, A., 1997, Participatory Ergonomics in a Red Meat Packing Plant, Part 1: Evidence of Long-Term Effectiveness, *American Industrial Hygeine Association Journal*, 58, 127-121.

Mumford, E. (1991), Participation in systems design - what can it offer? In *Human Factors for Informatics Usability* B. Shackel & S.J. Richardson, (eds.) Cambridge University Press, 267-290.

Nagamachi, M., 1991, Application of participatory ergonomics through quality-circle activities. In *Participatory Ergonomics*, Noro, K. and Imada, A.S. (eds.) pp. 139-165. Taylor & Francis, London.

Nagamachi, M., 1995, Requisites and practices of participatory ergonomics. *International Journal of Industrial Ergonomics, Special Issue: 'Participatory Ergonomics*, 15, 5, 371-379.

Nagamachi, M. and Tanaka, T., 1995, Participatory Ergonomics for Reengineering in a Chemical Fiber Company, Proceedings of the Human Factors and Ergonomics Society 39th Annual Meeting, 766-770.

Neumann, J., 1989, Why people don't participate when given the chance. *Industrial Participation*, No. 601 (Spring), 6-8.

Noro, K., 1991, Concepts, methods and people. In *Participatory Ergonomics*, Noro, K. and Imada, A.S. (eds.) pp. 3-30. Taylor & Francis, London.

O'Brien, D.D., 1981, Designing systems for new users. *Design Studies*, 2, 139-150.

O'Neill, M.J. and Robertson, M.M., 1996, A participatory design process for developing ergonomic office design criteria, *Human Factors in Organizational Design and Management*, 210-215.

Occupational Safety and Health Administration, (1991) Ergonomics Program . Management Guidelines for Meatpacking Plants (Reprint). OSHA Report 3123 US Department of Labor.

Oppenheim, A.N., 1992, *Questionnaire Design, Interviewing and Attitude Measurement* (2nd edition). Pinter Publishers, London.

Orifice, D., 1996, An Ergonomic Project in a Team Environment, *Human Factors in Organizational Design and Management*, Brown, Jnr., V.O. and Hendrick, H.W. (eds.) Elsevier Science B.V., 161-165.

Oxenburgh, M., 1991, *Increasing productivity and profit through health and safety*, CCH International, Chicago

Peters, R.H., 1989, Review of recent research on organizational and behavioral factors associated with mine safety. *Bureau of Mines Information Circular 9232*, U.S. Dept of Interior, Washington D.C.

Pikaar, R.N.,Thomassen, P.A.J., Degeling, P. and van Andel, H. 1990, Ergonomics in control room design, *Ergonomics, 33* (3) 589-600

Pransky, G., Snyder, T.B., Himmelstein, J., 1996, The Organizational Response: Influence on Cumulative Trauma disorders in the Workplace *In Beyond Biomechanics: Psychosocial Aspects of Musculoskeletal disorders in Office Work* Moon, S.D., and Sauter, S.L., (eds) Taylor and Francis, 251-262

Rawling, R.G., 1991, Participative approaches to the design of physical office environments. In *Participatory Ergonomics*, K. Noro and A. Imada (eds), Taylor and Francis, London. 53-72

Reuter, W. (1987. Procedures for participation in planning, developing and operating information systems. In *System Design for Human Development and Productivity: Participation and Beyond,* P. Doherty et al. (eds.) pp. 271-276. North-Holland, Amsterdam.

Ruohomaki,V., 1995, A simulation game for the development of administrative work processes. In *The Simulation and Gaming Yearbook,* Saunders, D. (ed.) Kogan Page, London. 264-270

Rouse, W.B. and Boff, K.R., 1997, Assessing Cost/Benefits of Human Fctors, In *Handbook of Human Factors and Ergonomics* (2nd edition) Salvendy, G., (ed) J. Wiley & Sons, Chichester 1617-1633

Salvendy, G., and Karwowski, W. (eds) 1994, *Design of Work and Development of Personnel in Advanced Manufacturing.* John Wiley & Sons, New York

Sanoff, H., 1985, The application of participatory methods in design and evaluation. *Design Studies,* 6(4), 178-180.

Saunders, D., 1995, Introducing simulations and games for business. In *The Simulation and Gaming Yearbook,* Saunders, D. (ed) Kogan Page, London 13-20.

Shipley, P., 1990, Participation ideology and methodology in ergonomics practice. In *Evaluation of Human Work: A Practical Ergonomics Methodology,* Wilson, J.R. and Corlett, E.N. (eds.) London: Taylor & Francis.

Silverstein, B.A., Richards, S.E., Alcser, K. and Schurman, S., 1991. Evaluation of in-plant ergonomics training. *International Journal of Industrial Ergonomics,* 8 179-193.

Simpson,G. and Mason, S.,1995, Economic analysis of ergonomics. In *Evaluation of Human Work: A Practical Ergonomics Methodology* 2nd Edition, Wilson, J.R. and Corlett, E.N. (eds.) Taylor & Francis, London.

Sinclair, M.A., 1995,Subjective Assessment. In *Evaluation of Human Work: A Practical Ergonomics Methodology,* 2nd Edition, J.R. Wilson & E.N. Corlett (eds.) Taylor & Francis, London. 69-100

Smith, M.J. and Zehel, D., 1992, Case Study No. 9, A stress reduction intervention programme for meat processors emphasizing job design and work organization (United States), *Conditions of Work Digest,* 11(2), 204-213.

Smith, J., 1994, A Case Study of a Participatory Ergonomics and Safety Program in a Meat Processing Plant, *Proceedings of the Twelfth Congress of the International Ergonomics Association* , 6(1), 114-116.

Smith, M.J. and Carayon-Sainfort, P., 1989, A balance theory of job design for stress reduction. *International Journal of Industrial Ergonomics,* 4 67-79

Snow, M.P., Kies, J.K., Neale, D.C. and Williges R.C., 1996, Participatory Design, *Ergonomics in Design*. 4(2), 18-24.

Soderberg, I., 1985, Office work and office automation influence over the work environment and work organization. Report 1985: 27, National Board of Occupational Safety and Health, Solna, Sweden.

St-Vincent, M., Fernandez, J., Kuorinka, I., Chicoine, D. and Beaugrand, S., forthcoming a, "Assimilation and use of ergonomic knowledge by non-ergonomists to improve jobs in two electrical product assembly plants". To be published in: *Human Factors and Ergonomics in Manufacturing*.

St-Vincent, M., Chicoine, D. and Beaugrand, S., forthcoming b, Validation of a Participatory Ergonomic Process in Two Plants in the Electrical Sector. To be published in: *International Journal of Industrial Ergonomics*.

Sullivan A., and Wilson, J.R., 1994, Ergonomics Design and Review Manual (restricted). Commonwealth Industrial Gases Ltd., Chatswood, NSW

Urlings, I.J.M. and Vink, P., 1995, Success and Failure in the Process Towards Workplace Improvement, *Advances in Industrial Egonomics and Safety VII*, Bittner, A.C. and Champney, P.C. (eds.) Taylor & Francis, 983-987.

U.S. Department of Labor, Occupational Safety and Health Administration (1990) Ergonomics Program Management Guidelines for Meatpacking Plants (DOL/OSHA Pub No 3123)

van der Schaaf, T.W. and Kragt, H. 1992, Redesigning a control room from an ergonomics point of view: a case study of user participation in a chemical plant In *Enhancing Industrial Performance*, H. Kragt (ed.), Taylor and Francis, London, pp 165-178

Vartiainen, M. and Ruohomäki, 1995, Group training with the teamwork game, In *The Simulation and Gaming Yearbook*, Saunders, D. (ed) Kogan Page, London 239-245.

Vartiainen, M. and Smeds, R., 1995, Preface to the Finland Symposium, In *The Simulation and Gaming Yearbook*, Saunders, D. (ed) Kogan Page, London 237-238.

Vink, P., Lourijsen, E., Wortel, E., Dul, J. 1992, Experiences in participatory ergonomics: results of a roundtable session during the 11th IEA Congress, Paris, July 1991. *Ergonomics* vol 35 no 2 123-127

Vink, P., Peeters, M., Grundemann, R.W.M., Smulders, P.G.W., Kompier, M.A.J. and Dul, J. 1995, A participatory approach to reduce mental and physical workload, *International Journal Industrial Ergonomics*, 15, 389-396

Vink, P. and Urlings, I.J.M., 1995, Success and failure factors in the process towards workplace improvement. In *Advances in Industrial Ergonomics and Safety VII*, Bittner, A.C. and Champney, P.C.(eds) Taylor & Francis

Vroom, V.H., 1964, *Work and Motivation.* John Wiley, Chichester.

Vroom, V.H. and Yetton, P.W. 1973, *Leadership and Decision Making* University of Pittsburgh Press, Philadelphia.

Wall, T.D., Corbett, J.M., Clegg, C.W., Jackson, P.R., and Martin, R., 1990, Advanced manufacturing technology: Towards a theoretical framework. *Journal of Organizational Behavior,* 11 201-219

Wagner, J.A., 1994, Participation's effect on performance and satisfaction: A reconsideration of research evidence. *Academy of Management Review,* 19 312-330

West, M., 1994, *Effective Teamwork.* The British Psychological Society, Leicester, UK.

Westlander, G., Viitasara, E., Johansson, A. and Shahnavaz, H., 1995, Evaluation of an ergonomics intervention programme in VDT workplaces. *Applied Ergonomics,* 26(2), 83-92.

Wilson, J.R., 1991a, Design Decision Groups - A participative process for developing workplaces. In *Participatory Ergonomics,* K. Noro and A. Imada (eds). Taylor and Francis, London.

Wilson, J.R. 1991b, A framework and a foundation for ergonomics?, *Journal of Occupational Psychology,* 64, 67-80.

Wilson, J.R., 1994, Devolving ergonomics: The key to ergonomics management programmes. *Ergonomics,* 37, 579-594.

Wilson, J.R. and Grey Taylor, S.M., 1995, Simultaneous engineering for self directed work teams implementation: A case study in the electronics industry. *International Journal of Industrial Ergonomics,* 16,

Wilson, J.R., 1995a, Ergonomics and Participation. In *Evaluation of Human Work: A Practical Ergonomics Methodology,* 2nd Edition, J.R. Wilson & E.N. Corlett (eds.) pp. 1071-1096. Taylor & Francis, London.

Wilson, J.R., 1995b, Solution ownership in participative work design: The case of a crane control room. *International Journal of Industrial Ergonomics,* 15, 329-344.

Wilson, J.R. and Haines, H.M., 1997, Participatory ergonomics, In *Handbook of Human Factors and Ergonomics* (2nd edition) Salvendy, G., (ed) J. Wiley & Sons, Chichester 490-513

Wilson, J.R., 1997, Virtual Environments and Ergonomics: Needs and Opportunities. To appear in *Ergonomics,* 40

Work Research Unit, 1982, *Meeting the Challenge of Change: Guidelines.* HMSO, London

Zink, K.J., 1994, Participatory Ergonomics: Some Examples in Occupational Safety, *Proceedings of the Twelfth Congress of the International Ergonomics Association*, 6(1), 106-108.

Zink, K.J., 1996, Continuous Improvement through Employee Participation. Some experiences from a long-term study in Germany, *Human Factors in Organizational Design and Management*, Brown, Jnr., V.O. and Hendrick, H.W. (eds.) Elsevier Science, 155-160.

Printed and published by the Health and Safety Executive
C3 6/98